Environmental Management Systems Handbook for Refineries

Environmental Management Systems Handbook for Refineries

Editor

Nitin Saini

scitus academics

Environmental Management Systems Handbook for Refineries
Edited by **Nitin Saini**

Printed in 2017

ISBN: 978-1-68117-367-2

Library of Congress Control Number: 2015941555

SCITUS Academics LLC,
616, Corporate Way, Suite 2, 4766,
Valley Cottage, NY 10989

www.scitusacademics.com

Notice

Reasonable efforts have been made to publish reliable data and views articulated in the chapters are those of the individual contributors, and not necessarily those of the editors or publishers. Editors or publishers are not responsible for the accuracy of the information in the published chapters or consequences of their use. The publisher believes no responsibility for any damage or grievance to the persons or property arising out of the use of any materials, instructions, methods or thoughts in the book. The editors and the publisher have attempted to trace the copyright holders of all material reproduced in this publication and apologize to copyright holders if permission has not been obtained. If any copyright holder has not been acknowledged, please write to us so we may rectify.

Contents

Preface

Refinery operations in complying with environmental laws and improving the overall efficiency of their operations. The author's approach has been recognized internationally as an integrated model that captures improved compliance and financial savings by reducing operating costs through dedicated pollution prevention programs. The focus is on refinery operations; however, the principles and general discussions are applicable to all manufacturing sectors. All business operations are faced with global challenges involving dynamic interactions between changing regulations, market competitiveness, and spiraling costs for raw materials and labor. It is impossible to separate environmental management issues and infra-structure investments from the primary strategies of any business. In fact, any company, facility, or refinery that focuses on pollution control as opposed to environmental management will not find itself in business over the next decade. There simply are too many costs and financial exposures to businesses that do not focus on continual improvement of their environmental performance. A commitment to continual improvement of environmental performance is increasingly becoming the standard of care for industry.

Editor

Chapter 1

Biodegradation: Involved Microorganisms and Genetically Engineered Microorganisms

Nezha Tahri Joutey[1], Wifak Bahafid[1], Hanane Sayel[1], and Naïma El Ghachtouli[1]

[1]Microbial Biotechnology Laboratory, Sidi Mohammed Ben Abdellah University, Faculty of Science and Technics, Fez, Morocco

INTRODUCTION

Biodegradation is defined as the biologically catalyzed reduction in complexity of chemical compounds [1]. Indeed, biodegradation is the process by which organic substances are broken down into smaller compounds by living microbial organisms [2]. When biodegradation is complete, the process is called "mineralization". However, in most cases the term biodegradation is generally used to describe almost any biologically mediated change in a substrate [3].

So, understanding the process of biodegradation requires an understanding of the microorganisms that make the process work. The microbial organisms transform the substance through metabolic or enzymatic processes. It is based on two processes: growth and cometabolism. In growth, an organic pollutant is used as sole source of carbon and energy. This process results in a complete degradation (mineralization) of organic pollutants. Cometabolism is defined as the metabolism of an organic compound in the presence of a growth substrate that is used as the primary carbon and energy source [4]. Several microorganisms, including fungi, bacteria and yeasts are involved in biodegradation process. Algae and protozoa reports are scanty regarding their involvement in biodegradation [5]. Biodegradation processes vary greatly, but frequently the final product of the degradation is carbon dioxide [6]. Organic material can be degraded aerobically, with oxygen, or anaerobically, without oxygen [4, 7].

Biodegradable matter is generally organic material such as plant and animal matter and other substances originating from living organisms, or artificial materials that are similar enough to plant and animal matter to be put to use by microorganisms. Some microorganisms have the astonishing, naturally occurring, microbial catabolic diversity to degrade, transform or accumulate a huge range of compounds including hydrocarbons (e.g. oil), polychlorinated biphenyls (PCBs), polyaromatic hydrocarbons (PAHs), radionuclides and metals [8].

The term biodegradation is often used in relation to ecology, waste management and mostly associated with environmental remediation (bioremediation) [2]. Bioremediation process can be divided into three phases or levels. First, through natural attenuation, contaminants are reduced by native microorganisms without any human augmentation. Second, biostimulation is employed where nutrients and oxygen are applied to the systems to improve their effectiveness and to accelerate biodegradation. Finally, during bioaugmentation, microorganisms are added to the systems. These supplemental organisms should be more efficient than native flora to degrade the target contaminant [9]. A feasible remedial technology requires microorganisms being capable of quick adaptation and efficient uses of pollutants of interest in a particular case in a reasonable period of time [10]. Many factors influence microorganisms to use pollutants as substrates or cometabolize them, like, the genetic potential and certain environmental factors such as temperature, pH, and available nitrogen and phosphorus sources, then, seem to determine the rate and the extent of degradation [4]. Therefore, applications of genetically engineered microorganisms (GEM) in bioremediation have received a great deal of attention. These GEM have higher degradative capacity and have been demonstrated successfully for the degradation of various pollutants under defined conditions. However, ecological and environmental concerns and regulatory constraints are major obstacles for testing GEM in the field [11].

In this chapter we will try to foster an in-depth understanding of biodegradation process by trying to cover all types of microorganisms implied in degradation of different pollutants. Moreover, although we are aware that the term biodegradation is often used in relation to ecology, waste management, biomedicine, and the natural environment (bioremediation) and is now commonly associated with environmentally friendly products, this chapter will mainly give attention to biodegradation in relation to bioremediation through describing processes (natural attenuation, biostimulation and bioaugmentation) utilizing degradation abilities of microorganisms in bioremediation and factors affecting this process. Microorganisms may be genetically engineered for many purposes. One such purpose is for the efficient degradation of pollutants. So, the second scope of this chapter is to demonstrate the importance of some GEM in this process and to describe obstacles for testing GEM in the field, which must be overcome before GEM can provide an effective clean-up process at lower cost. Figure 1 summarizes the contents of this chapter.

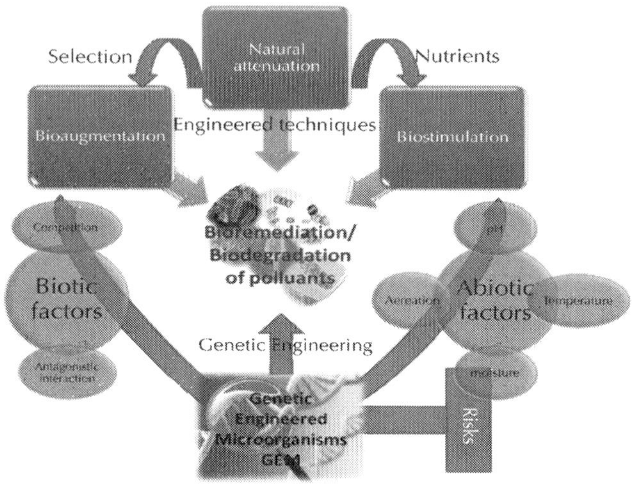

Figure 1: Bioremediation of polluants utilizing biodegradation abilities of microorganisms include the natural attenuation, although it may be enhanced by engineered techniques, either by addition of selected microorganisms (bioaugmentation) or by biostimulation, where nutrients are added. Genetic engineering is also used to improve the biodegradation capabilities of microorganisms by GEM. Nevertheless, there are many factors affecting the efficiency of this process and risks associated to the use of GEM in the field.

ROLE OF MICROORGANISMS IN BIODEGRADATION OF POLLUTANTS

In this chapter, biodegradation is described associated with environmental bioremediation. Therefore, biodegradation is nature's way of recycling wastes, or breaking down organic matter into nutrients that can be used and reused by other organisms. In the microbiological sense, "biodegradation" means that the decaying of all organic materials is carried out by a huge assortment of life forms comprising mainly bacteria, yeast and fungi, and possibly other organisms.

Bioremediation and biotransformation methods endeavour to harness the astonishing, naturally occurring, microbial catabolic diversity to degrade, transform or accumulate a huge range of compounds including hydrocarbons (e.g. oil), polychlorinated biphenyls (PCBs), polyaromatic hydrocarbons (PAHs), radionuclides and metals [12].

Some Biodegradable Pollutants

In the last few decades, highly toxic organic compounds have been synthesized and released into the environment for direct or indirect application over a long period of time. Fuels, polychlorinated biphenyls (PCBs), polycyclic aromatic hydrocarbons (PAHs), pesticides and dyes are some of these types of compounds [9]. Some other synthetic chemicals like radionuclides and metals are extremely resistant to biodegradation by native flora compared with the naturally occurring organic compounds that are readily degraded upon introduction into the environment.

- ***Hydrocarbons***: are organic compounds whose structures consist of hydrogen and carbon. Hydrocarbons can be seen as linear linked, branched or cyclic molecules. They are observed as aromatic or aliphatic hydrocarbons. The first one has benzene (C_6H_6) in its structure, while the aliphatic one is seen in three forms: alkanes, alkenes and alkynes [13].
- ***Polycyclic Aromatic Hydrocarbons (PAHs):*** are important pollutants class of hydrophobic organic contaminants (HOCs) widely found in air, soil and sediments. The major source of PAH pollution is industrial production [7]. They have been studied with increasing interest for more than twenty years because of more findings about their toxicity, environmental persistence and prevalence [14]. PAHs can sorb to organic-rich soils and sediments, accumulate in fish and other aquatic organisms, and may be transferred to humans through seafood consumption [7]. The biodegradation of PAHs can be considered on one hand to be part of the normal processes of

the carbon cycle, and on the other as the removal of man-made pollutants from the environment. The use of microorganisms for bioremediation of PAH-contaminated environments seems to be an attractive technology for restoration of polluted sites.

- ***Polychlorinated Biphenyls (PCBs):*** are mixtures of synthetic organic chemicals. Due to their non-flammability, chemical stability, high boiling point, and electrical insulating properties, PCBs were used in hundreds of industrial and commercial applications including electrical, heat transfer, and hydraulic equipment; as plasticizers in paints, plastics, and rubber products; in pigments, dyes, and carbonless copy paper; and many other industrial applications. Consequently, PCBs are toxic compounds that could act as endocrine disrupters and cause cancer. Therefore, environmental pollution with PCBs is of increasing concern [15].
- ***Pesticides:*** are substances or mixture of substances intended for preventing, destroying, repelling or mitigating any pest. Pesticides which are rapidly degraded are called nonpersistent while those which resist degradation are termed persistent. The most common type of degradation is carried out in the soil by microorganisms, especially fungi and bacteria that use pesticides as food source [16].
- ***Dyes:*** are widely used in the textile, rubber product, paper, printing, color photography, pharmaceuticals, cosetics and many other industries [17]. Azo dyes, which are aromatic compounds with one or more (–N=N–) groups, are the most important and largest class of synthetic dyes used in commercial applications [18]. These dyes are poorly biodegrabale because of their structures and treatment of wastewater containing dyes usually involves physical and / or chemical methods such as adsorption, coagulation-flocculation, oxidation, filtration and electrochemical methods [19].

The success of a biological process for color removal from a given effluent depends in part on the utilization of microorganisms that effectively decolorize synthetic dyes of different chemical structures.

- ***Radionuclides:*** a radionuclide is an atom with an unstable nucleus, characterized by excess energy available to be imparted either to a newly created radiation particle within the nucleus or via internal conversion. During this process, the radionuclide is said to undergo radioactive decay, resulting in the emission of gamma ray(s) and/or subatomic particles such as alpha or beta particles [20].
- ***Heavy metals:*** unlike organic contaminants, the metals cannot be destroyed, but must either be converted to a stable form or removed. Bioremediation of metals is achieved through biotransformation. Mechanisms by which microorganisms act on heavy metals include biosorption (metal sorption to cell surface by physicochemical mechanisms), bioleaching (heavy metal mobilization through the excretion of organic acids or methylation reactions), biomineralization (heavy metal immobilization through the formation of insoluble sulfides or polymeric complexes), intracellular accumulation, and enzyme-catalyzed transformation (redox reactions) [21]. The major microbial processes that influence the bioremediation of metals are summarized in Figure 2.

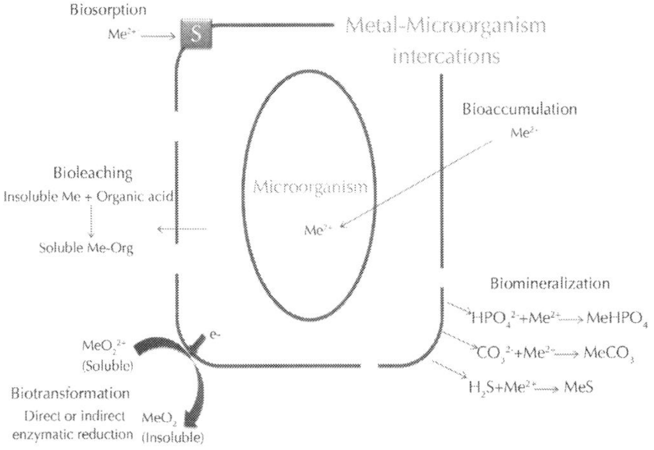

Figure 2: Microbial processes used in bioremediation technologies modified from Lloyd and Lovley [21].

Bacterial Degradation

There are many reports on the degradation of environmental pollutants by different bacteria. Several bacteria are even known to feed exclusively on hydrocarbons [22]. Bacteria with the ability to degrade hydrocarbons are named hydrocarbon-degrading bacteria. Biodegradation of hydrocarbons can occur under aerobic and anaerobic conditions, it is the case for the nitrate reducing bacterial strains*Pseudomonas* sp. and *Brevibacillus* sp. isolated from petroleum contaminated soil [23]. However, data presented by Wiedemeier et al. [24] suggest that the anaerobic biodegradation may be much more important. 25 genera of hydrocarbon degrading bacteria were isolated from marine environment [25]. Furthermore, among 80 bacterial strains isolated by Kafilzadeh et al. [26] which belonged to 10 genus as follows: *Bacillus, Corynebacterium, Staphylococcus, Streptococcus, Shigella, Alcaligenes, Acinetobacter, Escherichia, Klebsiella* and *Enterobacter, Bacillus* was the best hydrocarbon degrading bacteria.

Bacterial strains that are able to degrade aromatic hydrocarbons have been repeatedly isolated, mainly from soil. These are usually gram negative bacteria, most of them belong to the genus *Pseudomonas*. The biodegradative pathways have also been reported in bacteria from the genera *Mycobacterium, Corynebacterium, Aeromonas, Rhodococcus* and *Bacillus* [7].

Although many bacteria are able to metabolize organic pollutants, a single bacterium does not possess the enzymatic capability to degrade all or even most of the organic compounds in a polluted soil. Mixed microbial communities have the most powerful biodegradative potential because the genetic information of more than one organism is necessary to degrade the complex mixtures of organic compounds present in contaminated areas [27].

Both, anaerobic and aerobic bacteria are capable of biotransforming PCBs. Higher chlorinated PCBs are subjected to reductive dehalogenation by anaerobic microorganisms. Lower chlorinated biphenyls are oxidized by aerobic bacteria [28].

Research on aerobic bacteria isolated so far has mainly focused on Gram-negative strains belonging to the genera *Pseudomonas, Burkholderia, Ralstonia, Achromobacter, Sphingomonas* and *Comamonas*. However, several reports about PCB-degrading activity and characterization of the genes that are involved in PCB degradation indicated PCB-degrading potential of some Gram-positive strains as well (genera *Rhodococcus, Janibacter, Bacillus, Paenibacillus* and *Microbacterium*) [29]. Aerobic catabolic pathway for PCB degradation seems to be very similar for most of the bacteria and comprises four steps catalysed by the enzymes, biphenyl dioxygenase (BphA), dihydrodiol dehydrogenase (BphB), 2, 3-dihydroxybihenyl dioxygenase (DHBD) (BphC) and hydrolase (BphD) [30].

Successful removal of pesticides by the addition of bacteria had been reported earlier for many compounds, including atrazine [31]. Recent findings concerning pesticide degrading bacteria include the chlorpyrifos degrading bacterium *Providencia stuartii* isolated from agricultural soil [32] and isolates *Bacillus, Staphylococcus* and *Stenotrophomonas* from cultivated and uncultivated soil able to degrade dichlorodiphenyltrichloroethane (DDT) [33].

Researches on bacterial strains that are able to degrade azo dyes under aerobic and anaerobic conditions have been extensively reported [34]. Based on the available literature, it can be concluded that the microbial decolourization of azo dyes is more effective under anaerobic conditions. On the other hand, these conditions lead to aromatic amine formation, and these are mutagenic and toxic to humans requiring a subsequent oxidative (aerobic) stage for their degradation. In this context, the combined anaerobic/aerobic biological treatments of textile dye effluents using microbial consortia are common in the literature [35]. For exemple, Chaube et al. [36] have used the mix consortia of bacteria consisting of *Proteus* sp., *Pseudomonas* sp. and *Enterococcus* sp. in biodegradation and decolorisation of dye. However, several researchers have identified single bacterial strains that have very high efficacy for removal of azo dyes, it is the case of *Shewanella decolorations* [37]. In contrast to mixed cultures, the use of a pure culture has several advantages.

These include predictable performance and detailed knowledge on the degradation pathways with improved assurance that catabolism of the dyes will lead to nontoxic end products under a given set of environmental conditions. Another advantage is that the bacterial strains and their activity can be monitored using culture-based or molecular methods to quantify population densities of the bacteria over time. Knowledge of the population density can be extrapolated to quantitative analysis of the kinetics of azo dye decoloration and mineralization [38].

Heavy metals cannot be destroyed biologically (no "degradation", change in the nuclear structure of the element, occurs) but are only transformed from one oxidation state or organic complex to another [39]. Besides, bacteria are also efficient in heavy metals bioremediation. Microorganisms have developed the capabilities to protect themselves from heavy metal toxicity by various mechanisms, such as adsorption, uptake, methylation, oxidation and reduction. Reduction of metals can occur through dissimilatory metal reduction [40], where bacteria utilize metals as terminal electron acceptors for anaerobic respiration. In addition, bacteria may possess reduction mechanisms that are not coupled to respiration, but instead are thought to impart metal resistance. For example, reduction of Cr(VI) to Cr(III) under aerobic [41] or anaerobic conditions [42], reduction of Se(VI) to elemental Se [43], reduction of U(VI) to U(IV) [44] and reduction of Hg(II) to Hg(0) [45]. Microbial methylation plays an important role in heavy metals bioremediation, because methylated compounds are frequently volatile. For example, Mercury, Hg(II) can be biomethylated by a number of different bacterial species*Alcaligenes faecalis*, *Bacillus pumilus*, *Bacillus* sp., *P. aeruginosa* and *Brevibacterium iodinium* to gaseous methyl mercury [46]. In addition to redox conversions and methylation reactions, acidophilic iron bacteria like *Acidithiobacillus ferrooxidans* [47] and sulfur oxidizing bacteria [48] are able to leach high concentrations of As, Cd, Cu, Co and Zn from contaminated soils. On the other hand metals can be precipitated as insoluble sulfides indirectly by the metabolic activity of sulphate reducing bacteria [48]. Sulphate reducing bacteria are anaerobic heterotrophs utilizing a range of organic substrates with SO_4^{2-} as the terminal electron acceptor. Heavy metal ions can be entrapped in the cellular structure and subsequently biosorbed onto the binding sites present in the cellular structure. This method of uptake is independent of the biological metabolic cycle and is known as biosorption or passive uptake. The heavy metal can also pass into the cell across the cell membrane through the cell metabolic cycle. This mode of metal uptake is referred as active uptake. *Pseudomonas* strain, characterized as part of a project to develop a biosorbent for removal of toxic radionuclides from nuclear waste streams, was a potent accumulator of uranium (VI) and thorium (IV) [49].

Most works on pollutants bioremediation uses pure microbial cultures. However, the use of mixed microbial cultures is undoubtedly advantageous. Some of the best examples of enrichment cultures comprising several specific consortia involve the bioremediation. In the case of heavy metals removal, Adarsh et al. [50] have used an environmental bacterial consortium to remove Cd, Cr, Cu, Ni and Pb from a synthetic wastewater effluent. For Cr(VI) removal we reported that the survival and stability of bacteria are better when they are present as a mixed culture, especially, in highly contaminated areas and in the presence of more than one type of metal [51]. Indeed, the indigenous bacteria enriched from chromium contaminated biotopes, were able to remove Cr(VI) successfully in multi-contaminated heavy metal solution [51]. A microbial consortium consisting of three bacterial Pseudomonas species originally obtained from dye contaminated sites was capable of decolorizing textile effluent and dye faster than the individual bacteria under static conditions [52].

PGPR and PGPB Degradation

Plant associated bacteria, such as endophytic bacteria (non-pathogenic bacteria that occur naturally in plants) and rhizospheric bacteria (bacteria that live on and near the roots of plants), have been shown to contribute to biodegradation of toxic organic compounds in contaminated soil and could have potential for improving phytoremediation [53]. Plant growth promoting rhizobacteria (PGPR) are naturally occurring soil bacteria that aggressively colonize plant roots and benefit plants by providing growth promotion [54]. Some plants can release structural analogs of PAHs such as phenols, to promote the growth of hydrocarbon degrading microbes and their degradation on PAHs. For such plant/microbe systems, an important class of bacteria is *Pseudomonas* spp., have PGPR activity and hydrocarbon degrading capacity [55]. Furthermore, the rhizosphere of vegetation in contaminated field contains higher diversity of population of PAH-degrading bacteria, among which two *Lysini bacillus*strains were isolated [56]. Culturable PCB degraders were also associated with both the rhizosphere and root zone of mature trees growing naturally in a contaminated site, they were identified as members of the genus *Rhodococcus, Luteibacter* and *Williamsia,* which suggest that biostimulation through rhizoremediation is a promising strategy for enhancing PCB degradation in situ [57]. Also, the free living nitrogen fixer *Azospirillum lipoferum* generally found in the rhizoplane of the crop plants was used for Malathion degradation which is one of the largest organo phosphorus insecticides in the world [33]. Results from the literature suggest that heavy metals may be removed from contaminated soils using plant growth promoting rhizobacteria. The use of soil bacteria (often plant growth promoting bacteria (PGPB)) as adjuncts in metal phytoremediation can significantly facilitate the growth of plants in the presence of high (and otherwise inhibitory) levels of metals [58]. To increase the efficiency of contaminants extraction, it is interesting to apply plants combined to some microorganisms; such technique is called rhizoremediation [59].

Microfungi and Mycorrhiza Degradation

Microfungi are described as a group of organisms that constitute an extremely important and interesting group of eukaryotic, aerobic microbes ranging from the unicellular yeasts to the extensively

Environmental Factors

Soil type and soil organic matter content affect the potential for adsorption of an organic compound to the surface of a solid. Absorption is an analogous process wherein a contaminant penetrates into the bulk mass of the soil matrix. Both adsorption and absorption reduce the availability of the contaminant to most microorganisms and the rate at which the chemical is metabolized is proportionately reduced [126]. Variations in porosity of the unsaturated and saturated zones of the aquifer matrix may influence the movement of fluids and contaminant migration in groundwater. The ability of the matrix to transmit gases, such as oxygen, methane and carbon dioxide, is reduced in fine grained sediments and also when soils become more saturated with water. This can affect the rate and type of biodegradation taking place [126].The oxidation-reduction potential of a soil provides a measurement of the electron density of the system. Biological energy is obtained from the oxidation of compounds in which electrons are transferred to various more oxidized compounds referred to as electron acceptors. A low electron density (Eh greater than 50 mV) indicates oxidizing, aerobic conditions, whereas high electron density (Eh less than 50 mV) indicates reducing, anaerobic conditions [126].

DEGRADATION BY GENETICALLY ENGINEERED MICROORGANISMS

As mentioned above, bioaugmentation and biostimulation are methods that can be applied to accelerate the recovery of polluted sites. In the late 1970s and early 1980s, bacterial genes encoding catabolic enzymes for recalcitrant compounds started to be cloned and characterized. Soon, many microbiologists and molecular biologists realized the potential of genetic engineering for addressing biodegradation [128]. A genetically engineered microorganism (GEM) or modified microorganism (GMM) is a microorganism whose genetic material has been altered using genetic engineering techniques inspired by natural genetic exchange between microorganisms. These techniques are generally known as recombinant DNA technology. Genetically engineered microorganisms (GEMs) have shown potential for bioremediation of soil, groundwater and activated sludge, exhibiting the enhanced degrading capabilities of a wide range of chemical contaminants [129]. As soon as the prospect of releasing genetically modified microorganisms for bioremediation became a reality, much of the research effort in the field was aimed at biosafety and risk assessment [128].

There are at least four principal approaches to GEM development for bioremediation application [11]. These include: 1) Modification of enzyme specificity and affinity; 2) Pathway construction and regulation; 3) Bioprocess development, monitoring and control; 4) Bioaffinity bioreporter sensor applications for chemical sensing, toxicity reduction and end point analysis.

Genetically Engineered Microorganisms

Molecular biology offers the tools to optimize the biodegradative capacities of microorganisms, accelerate the evolution of "new" activities, and construct totally "new" pathways through the assemblage of catabolic segments from different microbes [130].

Genes responsible for degradation of environmental pollutants, for example, toluene, chlorobenzene acids, and other halogenated pesticides and toxic wastes have been identified. For every compound, one separate plasmid is required. It is not like that one plasmid can degrade all the toxic compounds of different groups. The plasmids are grouped into four categories: 1) OCT plasmid which degrades, octane, hexane and decane; 2) XYL plasmid which degrades xylene and toluenes, 3) CAM plasmid that decompose camphor and 4) NAH plasmid which degrades naphthalene [130].

The potential for creating, through genetic manipulation, microbial strains able to degrade a variety of different types of hydrocarbons has been demonstrated by Friello et al. [131]. They successfully produced a multiplasmid-containing *Pseudomonas* strain capable of oxidizing aliphatic, aromatic, terpenic and polyaromatic hydrocarbons.

Pseudomonas putida that contained the XYL and NAH plasmid as well as a hybrid plasmid derived by recombinating parts of CAM and OCT developed by conjugation could degrade camphor, octane, salicylate, and naphthalene [129] and could grew rapidly on crude oil because it was capable of metabolizing hydrocarbons more efficiently than any other single plasmid [131].

This product of genetic engineering was called as superbug (oil eating bug). The plasmids of *P. putida*degrading various chemical compounds are TOL (for toluene and xylene), RA500 (for 3, 5-xylene) pAC 25 (for 3-cne chlorobenxoate), pKF439 (for salicylate toluene). Plasmid WWO of *P. putida* is one member of a set of plasmids now termed as TOL plasmid. It was the first living being to be the subject of an intellectual property case. At that point, it seemed that molecular techniques, either through plasmid breeding or sheer genetic engineering, could rapidly produce microbes with higher catalytic abilities, able to basically degrade any environmental pollutant [129].

Reports on the degradation of environmental pollutants by genetically engineered microorganisms are focused on genetically engineered bacteria using different genetic engineering technologies: Pathway modification, modification of substrate specificity by *Comamonas testosteroni* VP44 [132].

The application of genetic engineering for heavy metals removal has aroused great interest. For example, *Alcaligenes eutrophus* AE104 (pEBZ141) was used for chromium removal from industrial wastewater [133] and the recombinant photosynthetic bacterium, *Rhodopseudomonas palustris,* was constructed to simultaneously express mercury transport system and metallothionein for Hg^{2+} removal from heavy metal wastewater [134].

For polychlorinated biphenyls degradation, chromosomally located PCB catabolic genes of *R. eutropha*A5, *Achromobacter* sp. LBS1C1, and *A. denitrificans* JB1 were transferred into a heavy metal resistant strain *R. eutropha* CH34 through natural conjugation [11].

Genetic engineering of endophytic and rhizospheric bacteria for use in plant associated degradation of toxic compounds in soil is considered one of the most promising new technologies for remediation of contaminated environmental sites [53]. To select a suitable strain for gene recombination and inoculation into the rhizosphere, there are three criteria that has been recommended: first, the strain should be stable after cloning and the target gene should have a high expression, second, the strain should be tolerant or insensitive to the contaminant; and third, some strains can survive only in several specific plant rhizosphere [135]. Many bacteria in the rhizosphere show only limited ability in degrading organic pollutants. With the development of molecular biology, the genetically engineered rhizobacteria with the contaminant-degrading gene are constructed to conduct the rhizoremediation [58]. Examples about the molecular mechanisms involved in the degradation of some pollutants such as trichloroethylene (TCE) and PCBs has been studied [136].

For heavy metals, Sriprang et al. [136] introduced *Arabidopsis thaliana* gene for phytochelatin synthase (PCS; PCSAt) into *Mesorhizobium huakuii* subsp. *rengei* strain B3 and then established the symbiosis between *M. huakuii* subsp. *rengei* strain B3 and *Astragalus sinicus*. The gene was expressed to produce phytochelatins and accumulate Cd^{2+}, under the control of bacteroid-specific promoter, the nifH gene [137].

Finally, the use of GEM strains as an inoculum during seeding would preclude the problems associated with competition between strains in a mixed culture. However, there is considerable controversy surrounding the release of such genetically engineered microorganisms into the environment, and field testing of these organisms must therefore be delayed until the issues of safety and the potential for ecological damage are resolved [138].

Obstacles Associated With the Use of Gem in Bioremediation Applications

While genetic engineering has produced numerous strains able to degrade otherwise intractable pollutants in a Petri dish or in a bioreactor, the practical translation of this research into actual in situ bioremediation practices has been quite scanty [129]. One major issue in this respect is the growing realization that the strains and bacterial species that most frequently appear in traditional enrichment procedures are not the ones performing the bulk of biodegradation in natural niches and may not even be any good as bioremediation mediators. The use of stable isotope probing (SIP) and equivalent methods in microbial ecology have revealed that *Pseudomonas, Rhodococcus*, and the typical aerobic fast growers that are widely favored as hosts of biodegradation related recombinant genes are far less significant under natural conditions [138]. Furthermore, using fast-growers as agents for biodegradation is the inevitable buildup of unwelcome biomass. As an alternative, the optimal clean-up agent would be the one that displays a maximum catalytic ability with a minimum of cell mass. The expression of biodegradation genes can be artificially uncoupled from growth with the use of stationary-phase promoters or starvation promoters [139]. In addition, the recent advances in the area of recombinant DNA technologies have paved the way for conceptualizing "suicidal genetically engineered microorganisms" (S-GEMS) to minimize such anticipated hazards and to achieve efficient and safer decontamination of polluted sites [140]. In some cases, whether the introduced bacterium is recombinant or not makes little difference, because the problem is that of implantation of foreign microbes in an unfamiliar territory. The introduction of bacterial biomass in an existing niche may create a palatable niche for protozoa that prevents the bacterial population to grow beyond a certain level [141]. Ingenious approaches have been developed to circumvent this problem, including encapsulation of the inoculum in a polymeric matrix or protection in plastic tubing [142].

The efficacy of a desired in-situ catalytic activity (biodegradation or otherwise) depends first on its presence in the target site. One key enzyme may not be there, or it may preexist in the site but not be manifested. Alternatively, it can be hosted by just a very minor part of the whole microbial population, so that its factual expression in the site might not be significant [129].

A field release of *P. fluorescens* HK44 for bioremediation application has been successfully conducted on moderately large-scale and controlled field condition [143]. However, the future application of genetically engineered bacteria for pollution remediation will not be free from the risks associated with their release in the environment. The future risk regarding use of other engineered bacteria is still unclear. Therefore, the future perspectives of engineered bacterial strains under the field conditions will be the focus of review, which may help us to assess the obstacles related with application of genetically engineered bacteria in environmental bioremediation.

The major problem encountered in successful bioremediation technology pertains to hostile field conditions for the engineered microbes. Besides, the molecular applications are mainly confined to only few well characterized bacteria such as *E. coli*, *P. putida*, *B. subtilis*, etc. Other bacterial strains need to be tried for developing the engineered microbes. The specific characteristic of open biotechnological applications has clearly necessitated the development of engineered bacterial strains to meet the new challenges. The main concern is to construct GE bacteria for field release in bioremediation with an adequate degree of environmental certainty. Efforts should be made to examine the performance of engineered bacteria in terms of their survival, potential of horizontal gene transfer, which may affect the indigenous microflora within a complex environmental situation. Often the novel scientific researches always give rise to still more fascinating questions pertaining to public concern. In the majority cases, the bacteria designed for bioremediation processes have been designed for specific purpose under the laboratory conditions, ignoring the field requirement and other complex situations. However, there is no evidence that the deliberate release of GE bacteria for bioremediation has caused a measurable adverse impact on the natural microbial community. At least the overstated idea of risk appraisal has fuelled so much debate and triggered so many research efforts, which have immensely contributed a lot in the field of environmental microbiology. However, survival of the GE bacteria in complex environmental situations is still a big question, needs to be addressed in the light of latest findings [144].

CONCLUSION

Microbial activities are very important for the renewal of our environment and maintenance of the global carbon cycle. These activities are included in the term biodegradation. Amid the substances that can be degraded or transformed by microorganisms are a huge number of synthetic compounds and other chemicals having ecotoxicological effects like hydrocarbons and heavy metals. However, in most cases this statement concerns potential degradabilities which were estimated in the laboratory by using selected cultures and under ideal growth conditions. Due to a whole range of factors: competition with microorganisms, insufficient supply with essential substrates, unfavorable external conditions (aeration, moisture, pH, temperature), and low bioavailability of the pollutant, biodegradation in natural conditions is lesser. So, environmental biotechnology has the objective of tackling and solving these problems so as to permit the use of microorganisms in bioremediation technologies. For this reason, it is necessary to support the activities of the indigenous microorganisms in polluted biotopes and to enhance their degradative abilities by bioaugmentation or biostimulation. Genetic engineering is also used to improve the biodegradation capabilities of microorganisms. Nevertheless, there are many risks associated to the use of GEM in the field. Whether or not such approaches are ultimately successful in bioremediation of pollutants may make a difference in our ability to reduce wastes, eliminate industrial pollution, and enjoy a more sustainable future.

ACKNOWLEDGEMENTS

The authors are grateful for the financial and scientific support rendered by Microbial Biotechnology Laboratory of Faculty of Sciences and Innovation City, SMBA University, Fez, Morocco.

REFERENCES

1. Alexander M. Biodegradation and Bioremediation, San Diego CA 1994. Academic Press.
2. Marinescu M, Dumitru M and Lacatusu A. Biodegradation of Petroleum Hydrocar- bons in an Artificial Polluted Soil. Research Journal of Agricultural Science 2009; 41(2).
3. Bennet JW, Wunch KG and Faison B D. Use of fungi biodegradation. Manual of envi- ronmental microbiology 2002, 2nd ed., ASM Press: Washington, D.C., 960-971.
4. Fritsche W and Hofrichter M. Aerobic Degradation by Microorganisms 2008 in Bio- technology Set, Second Edition (eds H.-J. Rehm and G. Reed), Wiley-VCH Verlag GmbH, Weinheim, Germany. doi: 10.1002/9783527620999.ch6m
5. Das N and Chandran P. Microbial Degradation of Petroleum Hydrocarbon Contami- nants: An Overview SAGE-Hindawi Access to Research Biotechnology. Research International. Volume 2011, Article ID 941810, 13 pages. Doi:10.4061/2011/941810.
6. Pramila R., Padmavathy K, Ramesh KV and Mahalakshmi K. Brevibacillus parabrevis, Acinetobacter baumannii and Pseudomonas citronellolis - Potential candidates for biodegradation of low density polyethylene (LDPE). Journal of Bacteriology Research 2012; 4(1) 9-14.
7. Mrozik A, Piotrowska-Segèt Z and Labuzek S. Bacterial degradation and bioremedia- tion of Polycyclic Aromatic Hydrocarbons. Polish Journal of Environmental Studies 2003; 12 (1) 15-25.

8. Leitão AL. Potential of Penicillium Species in the Bioremediation Field. International Journal of Environmental Resarch. Public Health 2009; 6 1393-1417; doi:10.3390/ ijerph6041393.
9. Diez MC. Biological Aspects involved in the degradation of organic pollutants. Jour- nal of Soil Science and Plant Nutrition. online. 2010; 10 (3) 244-267. ISSN 0718-9516.
10. Seo JS, Keum YS and Li QX. Bacterial Degradation of Aromatic Compounds. Interna- tional Journal of Environmental Resarch. Public Health 2009, 6, 278- 309; doi:10.3390/ ijerph6010278.
11. Menn FM, Easter JP and Sayler GS. Genetically Engineered Microorganisms and Bio- remediation, in Biotechnology: Environmental Processes II 2008; 11b, Second Edition (eds H.-J. Rehm and G. Reed), Wiley-VCH Verlag GmbH, Weinheim, Germany. doi: 10.1002/9783527620951.ch21.
12. Lesley A. Ogilvie and Penny R. Hirsch. 2012. Microbial ecological theory: Current perspectives.
13. McMurry J (2000). In Aromatic hydrocarbons. Organic Chmistry.5th ed brooks/cole: Thomas learning. New York, pp. 120-180.
14. Okere UV and Semple KT. Biodegradation of PAHs in 'Pristine' Soils from Different Climatic Regions. Okere and Semple. J BioremedBiodegrad 2012, S:1. http:// dx.doi. org/10.4172/2155-6199.S1-006. Biodegradation: Involved Microorganisms and Genetically Engineered Microorganisms http://dx.doi.org/10.5772/56194 309
15. Seeger M, Hernández M, Méndez V, Ponce B, Córdova1 M and González M. Bacteri- al degradation and bioremediation of chlorinated herbicides and biphenyls. Journal of Soil Science and Plant Nutrition 2010; 10 (3) 320-332.
16. Vargas JM. Pesticide degradation. Journal of Arboriculture 1975; 1(12) 232- 233.
17. Raffi F, Hall JD and Cernigila, CE. Mutagenicity of azo dyes used in foods, drugs and cosmetics before and after reduction

by Clostridium species from the human intesti- nal tract. Food and chemical Toxicology 1997; 35 897-901.

18. Vandevivere PC, Bianchi R, Verstraete W. Treatment and reuse of wastewater from the textile wet-processing industry: review of emerging technologies. Journal of Chemical Technology and Biotechnology 1998; 72 289-302.
19. Verma P, Madamwar, D. Decolorization of synthetic dyes by a newly isolated strain of Serratia marcescers. World Journal of Microbiology and Biotechnology 2003; 19 615- 618.
20. Petrucci RH, Harwood WS and Herring FG, General Chemistry (8th ed., PrenticeHall 2002), p.1025-26.
21. Lloyd JR and Lovley DR. Microbial detoxification of metals and radionuclides. Cur- rent Opinion in Biotechnology 2001; 12 248–253.
22. Yakimov MM, Timmis KN and Golyshin PN. Obligate oil-degrading marine bacteria. Current Opinion in Biotechnology, 2007; 18(3) 257-266.
23. Grishchenkov VG, Townsend RT, McDonald TJ, Autenrieth RL, Bonner JS, Boronin AM Degradation of petroleum hydrocarbons by facultative anaerobic bacteria under aerobic and anaerobic conditions. Process Biochemistry 2000; 35(9) 889-896.
24. Wiedemeier TH, Miller RN and Wilson JT. Significance of anaerobic processes for the Intrinsic bioremediation of fuel hydrocarbons: In, Proceedings of the Petroleum Hy- drocarbons and Organic Chemicals in Groundwater - Prevention, Detection, and Re- mediation Conference, November 29 - December 1, 1995, Houston Texas.
25. Floodgate G. The fate of petroleum in marine ecosystems. In Petroleum Microbiolo- gy, R. M. Atlas, Ed., pp. 355–398, Macmillion, New York, NY, USA, 1984.
26. Kafilzadeh F, Sahragard P, Jamali H and Tahery Y. Isolation and identification of hy- drocarbons degrading bacteria in soil around Shiraz Refinery. African Journal of Mi- crobiology Research 2011; 4(19) 3084-3089.

27. Fritsche W and Hofrichter M. Aerobic degradation of recalcitrant organic com- pounds by microorganisms, in environmental biotechnology: Concepts and applica- tions (eds H.-J. Jördening and J. Winter), Wiley-VCH Verlag GmbH & Co. KGaA, Weinheim, FRG. 2005 doi: 10.1002/3527604 286.ch7. 310 Biodegradation - Life of Science
28. Seeger M, Cámara B, Hofer B. Dehalogenation, denitration, dehydroxylation, and an- gular attack on substituted biphenyls and related compounds by a biphenyl dioxyge- nase. Journal of Bacteriology 2001; 183 3548-3555.
29. Petric I, Hrak D, Fingler S, Vonina E, Cetkovic H, Begonja A Kolar and UdikoviKoli N. Enrichment and characterization of PCB-Degrading bacteria as potential seed cul- tures for bioremediation of contaminated soil. Food Technology and Biotechnology 2007; 45(1)11-20.
30. Taguchi K, Motoyama M and Kudo T. PCB/biphenyl degradation gene cluster in Rhodococcus rhodochrousK37, is different from the well-known bph gene clusters in Rhodococcus sp. P6, RHA1, and TA42. RIKEN Review 2001; 42. Focused on Ecomolec- ular Science Research.
31. Struthers JK, Jayachandran K, Moorman TB. Biodegradation of atrazine by Agrobacte- rium radiobacter J14a and use of this strain in bioremediation of contaminated soil. Applied of Environonmental Microbiology 1998; 64 3368-3375.
32. Surekha Rani M, Lakshmi V, Suvarnalatha Devi KP, Jaya Madhuri R, Aruna S, Jyothi K, Narasimha G and Venkateswarlu K. Isolation and characterization of a chlorpyri- fos degrading bacterium from agricultural soil and its growth response. African Jour- nal of Microbiology Research 2008; (2) 026-031.
33. Kanade SN, Ade1 AB and Khilare VC. Malathion Degradation by Azospirillum lipofe- rum Beijerinck. Science Research Reporter 2012; 2(1) 94-103.
34. Dos Santos AB, Cervantes JF, Van Lier BJ. Review paper on current technologies for decolourisation of textile

wastewaters: perspectives for anaerobic biotechnology. Bioresource Technology 2007; 98 2369-2385.

35. Lodato A, Alfieri F, Olivieri G, Di Donato, Marzocchella A, Salatino A, P. Azo-dye conversion by means of Pseudomonas sp. OX1. Enzyme and Microbial Technology 2007; 41 646-652.
36. Chaube P, Indurkar H, Moghe S. Biodegradation and decolorisation of dye by mix consortia of bacteria and study of toxicity on Phaseolus mungo and Triticum aestivum. Asiatic Journal of Biotechnology Resources 2010; 01 45-56.
37. Hong Y, Xu M, Guo J et al. Respiration and growth of Shewanella decolorations S12 with an azo compound as the sole electron acceptor. Applied and Environmental Microbiology 2007; 73 64-72.
38. Khalid A, Arshad M, and Crowley D. Bioaugmentation of Azo Dyes. The Handbook of Environmental Chemistry 2010; 9.
39. Garbisu C, Alkorta I. Phytoextraction: A cost-effective plant based technology for the removal of metals from the environment. Bioresource Technology 2001; 77(3) 229-236. Biodegradation: Involved Microorganisms and Genetically Engineered Microorganisms http://dx.doi.org/10.5772/56194 311
40. Fern´andez PM, Martorell MM, Fari˜na JI, and Figueroa LIC., Removal Efficiency of Cr6+ by Indigenous Pichia sp. Isolated from Textile Factory Effluent. The Scientific World Journal. 2012, Article ID 708213, 6 pages doi:10.1100/2012/708213.
41. Sayel H, Bahafid W, Tahri-Joutey N, Derraz K, Fikri-Benbrahim K, Ibnsouda-Korai- chi S and El Ghachtouli N. Cr(VI) reduction by Enterococcus gallinarum isolated from tannery waste-contaminated soil. Annal of Microbiology. DOI 10.1007/s13213-011-0372-9.
42. Zhu W., Chai L., Ma Z., Wang Y., Xiao H. and Zhao K. Anaerobic reduction of hexa- valent chromium by bacterial cells of Achromobacter sp. StrainCh1. Microbiological Research 2008; 163 616-623.

43. Yee N, Ma J, Dalia A, Boonfueng T, and Kobayashi DY. Se(VI) Reduction and the Precipitation of Se(0) by the Facultative Bacterium Enterobacter cloacae SLD1a-1 Are Regulated by FNR. Applied and environmental microbiology 2007; 73(6) 1914-1920.
44. Gao W and Francis AJ. Reduction of Uranium(VI) to Uranium(IV) by Clostridia. Ap- plied and environmental microbiology 2008; 74(14) 4580-4584.
45. Brim H, McFarlan SC, Fredrickson JK, Minton KW, Zhai M, Wackett LP, Daly MJ. Engineering Deinococcusradiodurans for metal remediation in radioactive mixed waste environments. Nature Biotechnology 2000; 18(1) 85-90. doi:10.1038/71986. PMID 10625398.
46. De Jaysankar, Ramaiah N and Vardanyan L. Detoxification of toxic heavy metals by marine bacteria highly resistant to mercury. Marine Biotechnology 2008; 10(4) 471-477.
47. Takeuchi F and Sugio T. Volatilization and Recovery of Mercury from Mercury-Pol- luted Soils and Wastewaters Using Mercury-Resistant Acidithiobacillus ferrooxidans strains SUG 2-2 and MON-1.Environmental Sciences 2006; 13(6) 305-316.
48. White C, Sharman AK, Gadd GM: An integrated microbial process for the bioreme- diation of soil contaminated with toxic metals. Nature Biotechnology1998; 16 572-575.
49. PinakiSar, Sufia K. Kazy, S.F. D'Souza. Radionuclide remediation using a bacterial bi- osorbent. International Biodeterioration& Biodegradation 2004; 54(2-3) 193-202.
50. Adarsh VK, Madhusmita Mishra, SanhitaChowdhury, 2M. Sudarshan, A.R. Thakur and 1S. Ray Chaudhuri. Studies on Metal Microbe Interaction of Three Bacterial Iso- lates From East Calcutta Wetland OnLine Journal of Biological Sciences 2007; 7(2) 80-88.
51. Tahri-Joutey N, Bahafid W, Sayel H, El Abed S and El Ghachtouli N. Remediation of hexavalent chromium by consortia of indigenous bacteria from tannery waste-

conta- minated biotopes in Fez, Morocco. International Journal of Environmental Studies 2011; (686) 901-912. 312 Biodegradation - Life of Science

52. Jadhav JP, Kalyani DC, Telke AA, Phugare SS, Govindwar SP. Evaluation of the effi- cacy of a bacterial consortium for the removal of color, reduction of heavy metals, and toxicity from textile dye effluent. Bioresource Technology 2010; 101(1) 165-173.
53. Divya B and Deepak Kumar M. Plant–Microbe Interaction with Enhanced Bioreme- diation. Research Journal of BioTechnology. 2011; 6 (4).
54. Saharan BS, Nehra V. Plant Growth Promoting Rhizobacteria: A Critical Review Life Sciences and Medicine Research, Volume 2011: LSMR-21.
55. Hontzeas N, Zoidakis J and Glick BR. Expression and characterization of 1-aminocy- clopropane-1-carboxylate deaminase from the rhizobacterium Pseudomonas putida UW4: A key enzyme in bacterial plant growth promotion. Biochimica et Biophysica Acta 2004; 1703 11-19.
56. Ma B, Chen HH, He YandXu JM. Isolations and consortia of PAH-degrading bacteria from the rhizosphere of four crops in PAH-contaminated field. 2010 19th World Con- gress of Soil Science, Soil Solutions for a Changing World.
57. Leigh BM, Prouzova P, Mackova M, Macek T, Nagle DP and Fletcher Kinuthia Mwangi JS. Polychlorinated Biphenyl (PCB)-Degrading Bacteria Associated with Trees in a PCB-Contaminated Site. Applied and Environmental Microbiology 2006; 4(72) 2331-2342.
58. Glick BR. Using soil bacteria to facilitate phytoremediation. Biotechnology Advances. 2010; 28 367-374.
59. Jing Y, He Z, Yang X. Role of soil rhizobacteria in phytoremediation of heavy metal contaminated soils. Journal of Zhejiang University Sciences 2007; 8 192-207.

60. Rossman AY. Microfungi: Molds, mildews, rusts and smuts in our living resources. Washington, DC: U.S. Department of the Interior National Biological Service 1995.
61. Spellman F. R. Ecology for non ecologists. 2008. Page 176.
62. Matavuly MN and Molitoris HP. Marine fungi degraders of poly-3-hydroxyalka- noate based plastic materials. Proceedings for Natural Sciencespublished by Mati- caSrpska 2009; 116 253-265.
63. Khan AG. Mycorrhizoremediation an enhanced form of phytoremediation. Journal of Zhejiang University Science 2006 7(7) 503-514.
64. Mörtberg M, Neujahr HY, Uptake of phenol in Trichosporon cutaneum, Journal of Bac- teriology1985; 161 615-619.
65. Bartha R. Biotechnology of petroleum pollutant degradation, Microbial Ecology1986. 12 155-172.
66. De Cássia Miranda R, de Souza CS, de Barros Gomes E, Lovaglio RB, Lopes CE and de Fátima Vieira de Queiroz Sousa M. Biodegradation of Diesel Oil by Yeasts Isolat- Biodegradation: Involved Microorganisms and Genetically Engineered Microorganisms http://dx.doi.org/10.5772/56194 313 ed from the Vicinity of Suape Port in the State of Pernambuco –Brazil. Brazilian ar- chives of biology and technology 2007; 50(1)147-152.
67. Mucha K, Kwapisz E, Kucharska U and Okruszeki A. Mechanism of aniline degrada- tion by yeast strain Candida methanosorbosa BP-6. Polish journal of microbiology 2010; 59(4) 311-315.
68. Ijah UJJ. Studies on relative capabilities of bacterial and yeast isolates from tropical soil in degradating crude oil. Waste Manage. 1998; 18 293.
69. Fritsche W, Hofrichter M, Aerobic degradation by microorganisms, in: Biotechnolo- gy. Vol. 11b, Environmental Processes, (-Rehm, H.-J., Reed, G., Eds.), Wiley-VCH, Weinheim 2000, pp. 145–167.

70. Sasek V, Volfova O, Erbanova P, Vyas B R M, Matucha M. Degradation of PCBs by white rot fungi, methylotrophic and hydrocarbon utilizing yeasts and bacteria. Bio- technology Letters 1993; 15 521-526.
71. Eaton D C. Mineralization of polychlorinated biphenyls by Phanerochaete chryso- sporium: a ligninolytic fungus. Enzymeand Microbial Technology 1985; 7 194-196.
72. P. Cabras, M. Meloni, F. M. Pirisi, G. A. Farris and F. Fatichenti. Yeast and pesticide interaction during aerobic fermentation. Applied Microbiology and Biotechnology 1988; 29(2-3) 298-301, DOI:10.1007/BF01982920.
73. Wang J, Chen C. Biosorption of heavy metals by Saccharomyces cerevisiae: A Review. Biotechnology Advances 2006; 24: 427-451.
74. Bahafid W, Sayel H, Tahri-Joutey N and EL Ghachtouli N. Removal Mechanism of Hexavalent Chromium by a Novel Strain of Pichia anomala Isolated from Industrial Effluents of Fez (Morocco). Journal of Environmental Science and Engineering 2011; 5 980-991.
75. Bahafid W, Tahri-Joutey N, Sayel H, Iraqui-Houssaini M and El Ghachtouli N. 2012.Chromium adsorption by three yeast strains isolated from sediments in Moroc- co. DOI: 10.1080/01490451.2012.705228.
76. KsheminskaH P., Taras M. Honchar, Galyna Z. Gayda, Mykhailo V. Gonchar. Extracellular chromate-reducing activity of the yeast cultures. Central European Science Journals 2006; 1(1) 137-149.
77. Ksheminska H, Fedorovych D, Honchar T, Ivash M and Gonchar M. Yeast Tolerance to Chromium Depends on Extracellular Chromate Reduction and Cr(III) Chelation. Food Technology and Biotechnology 2008; 46 (4) 419-426.
78. SaiSubhashini S, Kaliappan S, Velan M. Removal of heavy metal from aqueous solu- tion using Schizosaccharomyces pombe in free and alginate immobilized cells. 2nd In- ternational

Conference on Environmental Science and Technology 2011; 6107-111. 314 Biodegradation - Life of Science

79. Abruscia C., Marquinaa D., Del Amob A., Catalina F. Biodegradation of cinemato- graphic gelatin emulsion by bacteria and filamentous fungi using indirect impedance technique. International Biodeterioration and Biodegradation 2007; 60 137-14.

80. Francesc X. Prenafeta-Boldu, Andrea Kuhn, Dion M. A. M. Luykx, HeidrunAnke, Jo- han W. van Groenestijn and Jan A. M. de Bont. Isolation and characterisation of fungi growing on volatile aromatic hydrocarbons as their sole carbon and energy source. Mycological Research 2001; 105(4) 477-484.

81. Steliga T. Role of fungi in biodegradation of Petroleum Hydrocarbons in Drill Waste. Polish Journal of Environmental Studies 2012; 21(2) 471-479.

82. Chaillan F, Le Flèche A, Bury E et al. Identification and biodegradation potential of tropical aerobic hydrocarbon degrading microorganisms. Research in Microbiology 2004; 155(7) 587-595.

83. Singh H, Mycoremediation: Fungal bioremediation, Wiley-Interscience, New York, NY, USA, 2006.

84. Tigini V, Di Toro S, Belardo A, Prigione V, Fava F and Varese GC. Identification and characterization of the active mycoflora of a consortium in the bioremediation of a soil historically contaminated by polychlorinated biphenyls (PCBs). 4th European bio- remediation conference.

85. Donnelly PK, Fletcher JS. PCB metabolism by ectomycorrhizal fungi. Bulletin of En- vironmental Contamination and Toxicology1995; 54 507-513.

86. Dmochewitz S, Ballschmiter K. Microbial transformation of technical mixtures of pol- ychlorinated biphenyls (PCB) by the fungus Aspergillus niger. Chemosphere 1988; 17 111-121.

87. Gadd GM. Interaction of fungi with toxic metals. New Phytologist 1993; 124 25-60.

88. Treen-Sears ME, Martin SM and Volesky B. Propagation of Rhiloprzs juvanicus bio- sorbent. Applied and Environmental Microbiology 198; 448 137-141.

89. Pal TK, Bhattacharyya S, Basumajumdar A. Cellular distribution of bioaccumulated toxic heavy metals in Aspergillus niger and Rhizopus arrhizus. International Journal of pharma and bio sciences 2010; 1(2) 1-6.

90. Gonzalez-Chavez MC, Carrillo-Gonzalez R, Wright SF, Nichols K. The role of gloma- lin, a protein produced by arbuscularmycorrhizal fungi, in sequestering potentially toxic elements. Environmental Pollution 2004; 130(3) 317-323. doi:10.1016/j.envpol. 2004.01.004.

91. Bumpus JA. Biodegradation of azo dyes by fungi. In: Arora, D.K. (Ed.), Fungal Bio- technology in Agricultural, Food and Environmental Applications. Marcel Dekker, New York, pp. 2004 457-480.

92. Chaplain V, Mamy L, Vieublé-Gonod L, Mougin C, Benoit P, Barriuso E and Nélieu S. Fate of Pesticides in Soils: Toward an Integrated Approach of Influential Factors, Biodegradation: Involved Microorganisms and Genetically Engineered Microorganisms http://dx.doi.org/10.5772/56194 315 Pesticides in the Modern World - Risks and Benefits, Margarita Stoytcheva (Ed.) 2011 ISBN: 978-953-307-458-0.

93. Walker JD, Colwell RR, Vaituzis Z and Meyer SA. Petroleum-degrading achlorophyl- lous alga Protothecazopfi. Nature (London) 1975; 254 423-424.

94. Cerniglia CE and Gibson DT. Metabolism of naphthalene by Cunninghamella elegans. Applied and Environmental Microbiology 1977; 34 363-370.

95. Stapleton RD and Jr.,V.P. Singh. Biotransformations: Bioremediation Technology for Health and Environmental protection. 2002.

96. Rogerson A, and Berger J. Effect of crude oil and petroleum-degrading microorgan- isms on the growth of freshwater and

soil protozoa. Journal of General Microbiology 1981; 124 53-59.

97. Bossert I, and Bartha R. 1984. The fate of petroleum in soil ecosystems, p. 434-476. In R. M. Atlas (ed.), Petroleum microbiology. Macmillan Publishing Co., New York.
98. X.-C. Wang and H.-M. Zhao. 2007. Uptake and Biodegradation of Polycyclic Aromat- ic Hydrocarbons by Marine Seaweed. Journal of Coastal Research, SI 50 (Proceedings of the 9th International Coastal Symposium), 1056 – 1061. Gold Coast, Australia, ISSN 0749.0208.
99. Kobayashi H and Rittman BE. Microbial removal of hazardous organic compounds. Environmental Science and Technology 1982; 16 170A-183A.
100. Jinqi I and Houtian O. Degradation of azo dyes by algae. Environmental Pollution 1992; 75, 273-278.
101. El-Sheekh MM, Gharieb MM and Abou-El-Souod GW. Biodegradation of dyes by some green algae and cyanobacteria. International biodeterioration and biodegrda- tion 2009; 63 699-704.
102. Dwivedi S. Bioremediation of heavy metal by algae: Current and future perspective. Journal of Advanced Laboratory Research in Biology 2012; 3(3).
103. Davis TA, Volesky B, Mucci A. A review of the biochemistry of heavy metal biosorp- tion brown algae. Water Research 2003;37:4311-30.
104. Peňa-Castro JM, Martínez-Jerónimo F, Esparza-García F, Caňizares-Villanueva RO. Heavy metals removal by the microalga Scenedesmus incrassatulus in continuous cul- tures. Bioresource Technology 2004; 94 219-222.
105. Akhtar N, Iqbal M, Iqbal ZS and Iqbal J. Biosorption characteristics of unicellular green alga Chlorella sorokiniana immobilized in loofa sponge for removal of Cr(III). Journal of Environmental Sciences 2008; 20 231-239. 316 Biodegradation - Life of Science

106. Mattison RG, Taki H, Harayama S. The soil flagellate Heteromita globosa accelerates bacterial degradation of alkylbenzenesthrough grazing and acetate excretion in batch culture.Microbial Ecology 2005; 49: 142-150.

107. Chen X., Liu M., Hu F., Mao X. and Li H. Contributions of soil micro-fauna (protozoa and nematodes) to rhizosphere ecological functions. ActaEcologicaSinica 2007; 27(8) 3132-3143.

108. Demnerova K, Mackova M, Spevakova, V, Beranova K, Kochankova L, Lovecka P, Ryslava, E, Macek T. Two approaches to biological decontamination of groundwater and soil polluted by aromatics characterization of microbial populations. Internation- al Microbiology 2005; 8 205-211.

109. Olaniran AO, Pillay D, Pillay B. Biostimulation and bioaugmentation enhances aero- bic biodegradation of dichloroethenes. Chemosphere 2006; 63 600-608.

110. Kaplan CW, Kitts CL. Bacterial succession in a petroleum land treatment unit. Ap- plied and Environmental Microbiology 2004; 70 1777-1786.

111. El Fantroussi S and Agathos SN. Is bioaugmentation a feasible strategy for pollutant removal and site remediation? Current Opinion in Microbiology 2005; 8 268-275.

112. El Fantroussi S, Belkacemi M, Top EM, Mahillon J, Naveau H, Agathos SN. Bioaug- mentation of a soil bioreactor designed for pilot-scale anaerobic bioremediation stud- ies. Environmental Science & Technology 1999; 33 2992-3001.

113. Army US. Interim army policy on natural attenuation for environmental restoration: Washington, DC, Department of the Army, Assistant Chief of Staff for Installation Management, DAIM-ED-R (200-1c), September 12, 1995.

114. EPA (U.S. Environmental Protection Agency). 2001. A Citizen's Guide to Monitored Natural Attenuation. EPA 542-F-01-004, U.S. Environmental Protection Agency, Washington, D.C.

115. Li CH, Wong YS, Tam NF. Anaerobic biodegradation of polycyclic aromatic hydro- carbons with amendment of

iron(III) in mangrove sediment slurry. Bioresource Technology 2010; 101 8083–8092.

116. Shan HF, Kurtz HD, Freedman D L. Evaluation of strategies for anaerobic bioreme- diatidon of high concentrations of halomethanes. Water Research 2010; 44 1317-1328.

117. Bond DR, Lovley DR. Reduction of Fe(III) oxide by methanogens in the presence and absence of extracellular quinines. Environmental Microbiology 2002; 4 115-124.

118. Chen M., Shih K., Hu M., Li F., Liu C., Wu W. and Tong H. Biostimulation of Indige- nous Microbial Communities for Anaerobic transformation of pentachlorophenol in paddy soils of southern China. Journal of Agricultural and Food Chemistry 2012; 60 2967-2975. Biodegradation: Involved Microorganisms and Genetically Engineered Microorganisms http://dx.doi.org/10.5772/56194 317

119. Chen BY., Chen WM. nd Chang JS. Optimal biostimulation strategy for phenol deg- radation withindigenous rhizobium Ralstonia taiwanensis. Journal of Hazardous Ma- terials; 2007 B139 232-237.

120. Liliane RR, Meneghetti, AntônioThomé, Fernando Schnaid, Pedro D.M. Prietto and Gabriel Cavelhão. Natural Attenuation and Biostimulation of Biodiesel Contaminat- ed Soils from Southern Brazil with Different Particle Sizes. Journal of Environmental Science and Engineering 2012; B1 155-162.

121. Fatima M. Bento, Fla´vio A.O. Camargo, Benedict C. Okeke, William T. Frankenberg- er. Comparative bioremediation of soils contaminated with diesel oil by natural at- tenuation, biostimulation and bioaugmentation. BioresourceTechnology 2005; 96 1049-1055.

122. Yu KSH, Wong AHY, Yau KWY, Wong YS, Tam NFY. Natural attenuation, biostimu- lation and bioaugmentation on biodegradation of polycyclic aromatic hydrocarbons (PAHs) in mangrove sediments. Marine Pollution Bulletin 2005; 51(8-12)1071-1077.

123. Thierry L., Armelle B. and Karine J. Performance of bioaugmentation-assisted phy- toextraction applied to metal contaminated soils: A review . Environmental Pollution 2008; 153 497-522.
124. Mrozik A, Piotrowska-Seget Z. Bioaugmentation as a strategy for cleaning up of soils contaminated with aromatic compounds. Microbial Research 2006; (2009), doi: 10.1016/j.micres.2009.08.001.
125. Riser-Roberts E. Remediation of petroleum contaminated soils. Biological, physical, and chemical process. Lewis Publishers Inc (ed). CRC Press LLC. USA. 1998 1-542.
126. ERD: Environmental Response Division. 1998. Fundamental principles of bioreme- diation (An aid to the development of bioremediation proposals).
127. van der Heul RM. Environmental Degradation of petroleum hydrocarbons. 2009
128. Cases I, de Lorenzo V. Genetically modified organisms for the environment: stories of success and failure and what we have learned from them. International microbiol- ogy 2005; 8 213-222.
129. Sayler GS, Ripp S. Field applications of genetically engineered microorganisms for bioremediation processes. Current Opinion in Biotechnology 2000; 11 286-289.
130. Ramos JL, Díaz E, Dowling D, de Lorenzo V, Molin S, O'Gara F, Ramos C, Timmis KN. The behavior of bacteria designed for biodegradation. Biotechnology (N Y); 1994 12(13) 1349-56.
131. Markandey DK and Rajvaidya N. Environmental Biotechnology, 1st edition, APH Publishing corporation 2004; 79. 318 Biodegradation - Life of Science
132. Hrywna Y, Tsoi TV, Maltseva OV, Quensen JF, and Tiedje JM. Construction and characterization of two recombinant bacteria that grow on ortho- and para-substitut- ed chlorobiphenyls. Applied and Environmental Microbiology 1999; 65(5) 2163-2169.

133. Srivastava NK, Jha MK, Mall ID, Singh D. Application of Genetic Engineering for Chromium Removal from Industrial Wastewater. International Journal of Chemical and Biological Engineering 2010; 3 3.
134. Xu D and Pei J. Construction and characterization of a photosynthetic bacterium ge- netically engineered for Hg2+ uptake. Bioresource Technology 2011; 102 3083-3088.
135. Huang XD, El-Alawi Y, Penrose DM, Glick BR, Greenberg BM. Responses of three grass species to creosote during phytoremediation. Environmental Pollution 2004; 130 453-63.
136. Sriprang R, Hayashi M, Ono H, Takagi M, Hirata K, Murooka Y. Enhanced accumu- lation of Cd2+ by a Mesorhizobium sp. transformed with a gene from Arabidopsis thaliana coding for phytochelatin synthase. Applied and Environmental Microbiolo- gy 2003; 69: 1791-1796.
137. Sussman MC, Collins H, Skinner FA, and Stewart-Tull DE (ed.). 1988. Release of ge- netically-engineered micro-organisms. Academic Press, Inc. (London), Ltd., London.
138. Wackett LP. Stable isotope probing in biodegradation research. Trends Biotechnol 2004; 22:153-154.
139. Matin A. Starvation promoters of Escherichia coli. Their function, regulation, and use in bioprocessing and bioremediation. Annals of the New York Academy of Sciences 1994; 721: 277-291.
140. Pandey G, Paul D, Jain RK. Conceptualizing "suicidal genetically engineered micro- organisms" for bioremediation applications. Biochemical and Biophysical Research Communications 2005; 327 637-639.
141. Iwasaki K, Uchiyama H, Yagi O. Survival and impact of genetically engineered Pseu- domonas putida harboring mercury resistance gene in aquatic microcosms. Bioscience Biotechnology & Biochemistry1993; 57 1264-1269.
142. Foster L, John R, Kwan Boon N, De Gelder L, Lievens H, Siciliano SD, Top EM, Ver- straete W. Bioaugmenting

bioreactors for the continuous removal of 3-chloroaniline by a slow release approach. Environmental Science & Technology 2002; 36 4698-46704.

143. Ripp S, Nivens DE, Ahn Y, Werner C, Jarrel J, Easter JP, Cox CD, Burlage RS, Sayler GS: Controlled field release of a bioluminescent genetically engineered microorganism for bioremediation process monitoring and control. Environmental Science & Technology 2000; 34 846-853. Biodegradation: Involved Microorganisms and Genetically Engineered Microorganisms http://dx.doi.org/10.5772/56194 319

144. Singh JS, Abhilash PC, Singh HB, Singh RP, Singh DP. Genetically engineered bacte- ria: An emerging tool for environmental remediation and future research perspectives. Gene 2011; 480 1-9.

Introduction of Environmental Management System (EMS)	Establishment of policies or goals for the EMS on a voluntary basis, introduce the EMS to facilities, and manage operations based on the EMS	(i) Acquisition of ISO14001 or other ESM certification by the third party organization, or introduction of a similar management system and its continuous implementation (ii) Ensuring commitment from top management
Identification of hazards or risk	Identify hazards or risks to the environment	Utilization of Material Safety Data Sheet (MSDS) or laboratory results identifying the risks or hazards in facilities (hazardous substances or appropriate packaging, etc.)
Occupational safety and health	Avoid exposures to unacceptable occupational risks in working environments	Ensuring occupational health and safety of employees (providing equipment for personal protection, cleaning of the working place and sanitary management, periodical medical checks, etc.)
Awareness and competency of staff	Enable workers to understand the hazards or risks of hazardous waste and handle it in an appropriate manner	Awareness raising and capacity development of employees (through seminars or in-house training, etc.)
Monitoring, recording, reporting	Collect and maintain records of information such as the material flow or emission status (exhaust gas or effluent) and report information to the appropriate authority	(i) Development of procedures for monitoring and allocating staff and equipment for monitoring (ii) Development of procedures for record keeping (iii) Periodical reporting to the appropriate authority

Emergency response	Capable of considering unexpected situations that can create negative effects on the environment	(i) Development of an emergency response plan (ii) Sharing of the above plan among facility employees and local authorities
Secure financial resource	Access to a financially stable resource that can accommodate unexpected situations (such as an accident or closure of the facility)	Ensuring access to a financial resource or having financial instruments (e.g., insurance)
Ensure downstream ESM	Ensure ESM in downstream recycling chain or destination country	(i) Maintaining contracts with recyclers in the downstream chain (ii) Maintaining contracts with exporters and all shipping documents

Environmental Quality (Scheduled Wastes) Regulations 2005 of Malaysia

In Malaysia, the release of the Environmental Quality (Scheduled Wastes) Regulations 2005, which replaced the revoked Environmental Quality (Scheduled Wastes) Regulations 1989, has classified e-waste as a scheduled waste under the code SW110, while the specified e-waste, such as waste from lead acid batteries, batteries containing heavy metals and fluorescent lamps, are coded as SW102, SW103, and SW109, respectively.

Because of this classification, any treatment of e-waste is regulated and must be carried out at a licensed on-site treatment facility, and the disposal of e-waste must be performed at the only prescribed premise, Kualiti Alam Sdn. Bhd. Because Malaysia is one of the parties to the Basel Convention, the export or import of e-waste is strictly prohibited, which is stipulated in Article 4, paragraph 1(a) and 1(b) in the Basel Convention on the Control of Transboundary Movements of Hazardous Wastes and their Disposal:

- Parties (refer to the country) exercising their right to prohibit the import of hazardous waste or other waste for disposal shall inform the other Parties of their decision pursuant to Article 13 (Transmission of Information);
- Parties shall prohibit or shall not permit the export of hazardous waste and other wastes to the Parties that have prohibited the import of such waste, when notified pursuant to subparagraph (a) above.

The Basel Convention on the Control of Transboundary Movements of Hazardous Wastes and their Disposal lists hazardous waste, mentioned in paragraph 1(a) of Article 1, as waste that belongs to any category contained in Annex I (Categories of Wastes to be Controlled), unless they do not possess any of the characteristics contained in Annex III (List of Hazardous Characteristics). Wastes listed in Annex VIII are also characterized as hazardous pursuant to paragraph 1(a) of Article 1.

ASSESSMENT OF E-WASTE SCENARIOS

e-Waste Management in Developing and Developed Countries

In India, e-waste is a major issue resulting from the generation of domestic e-waste and from waste that is imported from developed countries. India's electronics industry is one of the fastest growing industries in the world [13]. In 2005, India's Central Pollution Control Board developed guidelines for the environmentally sound management of e-waste in India; however, e-waste is not regulated at the present time. The Ministry of Environment and Forest, as part of the Environmental Protection Act of India, has enacted the "e-waste (Management and Handling) Rule" of 2011 that was to take effect on May 1, 2012.

The rule mandates producers to be responsible for the collection of e-waste and the financing of collection systems according to an extended producer responsibility concept. The rule clearly defines the responsibilities of the producers, collection centers, consumers, or bulk consumers, dismantlers and recyclers [14]. e-Waste dealers, refurbishers, dismantlers, recyclers, and collection centers are required to register with the relevant State Pollution Control Board (SPCB) or Pollution Control Committee (PCC). The liability on the consumer has also been fixed in the form of a requirement to dispose of e-waste by taking it to authorized dealers and collection centers.

Bulk consumers are at liberty to auction their waste with the restriction that they may only auction it to authorized collection centers, dismantlers, or recyclers or to the collection services offered by the producers. The rules assign all responsibility for ensuring enforcement to the respective SPCB or PCC. Every institution registered by the authorities has to submit an annual report to the appropriate SPCB or PCC [15].

In 2009, Thailand generated approximately 80,000 tons/year of e-waste that contained approximately 20,000 tons/year of electrical and electronic goods. Thailand also suffers from issues stemming from a lack of general awareness of e-waste, incomplete databases and inventories related to e-waste, a lack of environmentally sound management practices, and a lack of specific laws and regulations for e-waste. To address these issues, the government of Thailand passed the National Strategic Plan on Integrated Management of WEEE (WEEE Strategic Plan) in July 2007. The strategy, which was approved by the Cabinet on 24 July 2007, is a 10-year road map [16].

Japan has formulated two laws concerning e-waste production in Japan: the Law for the Promotion of Effective Utilization of Resources (LPUR) and the Law for Recycling Specified Kinds of Home Appliances (LRHA). The LPUR covers personal computers and small-sized batteries, whereas the LRHA is concerned with televisions, refrigerators, and washing machines, air conditioners, and clothes dryers. While the LPUR encourages manufacturers to voluntarily assist e-waste recycling to reduce the generation of waste, the LRHA imposes more compulsory obligations on the consumers and manufacturers. When disposing of home appliances, consumers are required to pay for the cost of transportation and recycling [15].

Manufacturers are responsible for establishing proper e-waste recycling facilities and are required to achieve compulsory recycling rates that include 70% for air conditioners, 60% for refrigerators, and 65% for washing machines. An amendment to the Law for the Effective Utilization of Resources took place on July 1, 2006, that mandates that manufacturers provide material content declarations for certain categories of electronic products sold after July 1, 2006. Manufacturers and importers are required to label their products and provide information on the content of lead, mercury, chromium VI, cadmium, Polybrominated biphenyls (PBB), and Polybrominated Diphenyl Ethers (PBDE). The importers of electronic products must also meet the Design for Environment (DfE) criteria, which are also required for domestic manufacturers [16].

In the 1980s, the policy in Germany of incinerating municipal solid waste faced increasing public opposition as waste volumes continued to grow to a point where the country was facing a waste disposal crisis. To overcome these challenges, the German government adopted the Packaging Ordinance following consultations with industry and consumers. This ordinance imposed significant packaging take-back and recycling obligations on the industry. Specifically, the ordinance established a requirement for retailers, rather than producers, to either take back packaging from consumers at retail shops or participate in a national collective system for packaging collection and recycling. Recognizing the significant challenges associated with collecting used packaging at retail shops, the retail, consumer goods and packaging industry established a voluntary organization—the "Duales System Deutschland" (DSD)—to collectively conduct their joint packaging management responsibilities under the ordinance [17].

The DSD organization established a packaging recycling and collection system across all of Germany that operated in parallel with the municipal waste collection system, hence its name meaning the "dual" system. Collection and processing from households and small businesses is physically undertaken by service providers operating under contract to the DSD, and the recycling of used packaging is provided by the DSD without directly charging consumers. However, to finance their activities, the DSD charges license fees to producers that sell package goods (in practice, packers and fillers) using the well-known "Green Dot" trademark as a way of proving that appropriate license fees have been paid [16].

e-Waste Management Issues in Malaysia

e-Waste in Malaysia is handled by the licensed recovery facilities. However, there are issues faced by the recovery facilities in achieving the goal of converting e-waste into a source material. From the visit and survey made to these facilities, the common issues are summarized as below.

e-Waste Supply

The issue of lacking e-waste supply is brought up at all full recyclers visited. This seems to be the main challenge faced by all large plants, as it relates to production of the plants, thus income to the plants. In Malaysia, many full recyclers are not able to operate at full capacity due to lack of e-waste. However, the recyclers have been diligent in investing, purchasing only the necessary shredder and grinder, and have been able to recover their investment costs. Also competition has led to stiff competition during bidding for wastes. Many full recyclers spend an inordinate amount of time and resources preparing for bids.

The recyclers alleged that there are too many full recyclers. Some allege that the recycling licenses have been issued without considering the amount of e-waste available. Also there are allegations that the e-waste has been illegally processed by illegal backyard operators without proper facilities. A few of the operators allege that much e-waste has been exported, with lack of checks at the customs attributing to e-waste getting through.

Importation of e-Waste Derived Products and Coding

Any full recyclers hold contract for e-waste processing for manufacturers with plants located in other countries, such as Campuchea, where labour is still cheap. In such a situation if parts from the waste are no longer coded as waste, but as metal scrap and metal mixture, there would be greater opportunity for full recyclers to grow.

Ingots lower in gold content are currently exported by certain full recyclers for refining, for example, to Japan. For equipment, such as a washing machine or a printer, only a small weight fraction of it is the electronic part comprising mainly the motherboard, plus several connectors, and so forth. Only the electronic part contains the precious metals recovered by e-waste recyclers, the rest are metal and plastic casings, wires, and so forth. Partial recyclers would isolate the electronic parts and, rightly, sell these to full recyclers for processing in physic-chemical-thermal processes. As it is now, the whole equipment is classified as SW110 Waste and the electronic parts remain as SW110 waste. There is need for greater clarity on coding, as the metal and plastic parts are no longer wastes but become feed for metal smelters and plastic recyclers, while the electronic parts become feed for physic-chemical-thermal processes.

Criteria for e-Waste Processing Technologies

In Malaysia, e-waste is collected and treated only at licensed facilities by the Department of Environment. By the end of February 2013, 18 full and 128 partial recovery facilities of e-waste had been licensed to perform the recovery process of precious (gold, silver, and platinum) and valuable (copper, aluminum, and nickel) metals. These facilities are also expected to simultaneously handle the hazardous elements found in e-waste, such as heavy metals (cadmium, lead, mercury), to prevent them from polluting the environment and risking human health and manage the recyclable items, such as plastic and glass.

Wastewater and sludge are primarily generated from processes of chemical dissolution, electrowinning, electrolysis, precipitation, and scrubbers. Sludge generated from wastewater treatment plants at the e-waste recovery facilities is categorized as a scheduled waste and coded SW501. The generated sludges are then sent to Kualiti Alam for proper disposal. The residue gas is treated using either a dust collector (mainly for the dust from crushing and grinding processes) or gas scrubber (vapor/gas from stripping and refining processes).

The existing technologies in use are conventional dismantling, crushing, grinding, air separation, or wet separation via jigging, chemical extraction, electrowinning, and thermal refining. Most equipment is similar to conventional mineral extraction equipment used in the mining industry and this equipment can be bought or fabricated in the country. Crushing, grinding, and air separation can be noisy and dusty, although there are technologies which are less noisy and dusty in use.

Chemical and thermal processes also pose a significant threat. Chemical extraction processes in use can emit chemical odors (usually of acids) although there are technologies with efficient vapour extraction and with cyanide detectors, whilst thermal refining processes are usually in open spaces, which may pose danger of metal vapours to workers, notably of lead, mercury, and so forth.

Destruction of, and metal recovery from, carbon ray tubes is a residual problem, as the tubes contain mercury vapours and yield leaded glass for which there is scarce demand; yet the carbon ray tubes are no longer in production. There remain only residual carbon ray tubes to be destroyed.

Since chip technologies are changing into those with smaller sizes, with thinner layers of precious metals, as such, the conventional physical separation processes of grinding or jigging may no longer be able to recover the precious metals.

Therefore, there is a need to categorise the technologies in use according to criteria of environmental performance, technological performance, and economics.

Certifications

To meet international standards for the e-waste industry, most e-waste recovery facilities have applied for certification through an accreditation process such as from the International Organization for Standardization (ISO). Certification includes requirements for quality, the environment and the occupational health and safety management system. Table 4 lists several types of certification that have been awarded to licensed e-waste recovery facilities.

Table 4: Types of certification awarded to e-waste recovery facilities in Malaysia

Number	Recovery facility	ISO 9001:2008	ISO 14001:2004	Responsible recycle (R2) practice	OHSAS 18001:2007	Others
1	A	—	Yes	—		—
2	B	—	Yes	—		—
3	C	Yes	Yes	—		—
4	D	—	Yes	—		—
5	E	Yes	Yes	In progress		—
6	F	—	Yes	—		—
7	G	Yes	Yes	—		—
8	H	Yes	Yes	—		—
9	I	Yes	Yes	—	Yes	—
10	J	Yes	Yes	—	—	—
11	K	Yes	Yes	Yes	—	—
12	L	Yes	Yes	—	—	—

13	M	Yes	Yes	—	Yes	MS 1722
14	N	Yes	Yes	—	—	—
15	O	—	—	—	—	—
16	P	—	—	—	—	—

ISO 9001:2008 sets out the criteria for a quality management system. The standard is based on a number of quality management principles that incorporate a strong customer focus, motivation and action of top management, the process approach, and continual improvement. By having this certificate, the facilities ensure that customers will receive consistent, good quality products and services, which can prove beneficial to business. ISO 14001:2004 sets out the criteria for an environmental management system and requires e-waste facilities to map out a framework for an effective environmental management system. These criteria can also provide assurance to facility management and employees, as well as to external stakeholders, that the facility's environmental impact is being measured and improved.

Some facilities were also awarded the International Standard (OHSAS 18001:2007) or Malaysia Standard (MS 1722) for occupational health and safety management systems. These facilities can use the awards to demonstrate their commitment to safeguarding the welfare of employees and others from workplace injuries and illnesses, as well as to meeting OSH legal requirements. One facility was awarded the Responsible Recycling Practices Standard (R2) certification from a third-party agency; this certification is recognized by the Environmental Protection Agency (EPA) and it includes a set of voluntary principles and guidelines designed to promote and assess responsible practices for electronics recyclers. The management system is accountable for practices affecting worker health and safety, security, the environment, and the downstream management, both domestically and internationally, of end-of-life electronic material and equipment.

CONCLUSIONS

Malaysia has adopted measures such as collecting information about the generation of e-waste, providing regulations and measures, and permitting only licensed facilities to perform treatment and recovery of e-waste. However, more drastic measures are required to ensure that all e-waste is collected, treated, and recovered by licensed facilities. Criteria should also be established mandating that the technologies are environmentally sound and do not produce deleterious effects to the people involved in handling the e-waste. In summary, the criteria for Environmentally Sound Management (ESM) must be provided and implemented for use at all levels of the management of e-waste.

ACKNOWLEDGMENTS

The authors would like to thank the Hazardous Waste Management Section of the Department of Environment (DOE) in Malaysia for their support of this study.

REFERENCES

1. D. Mohan and P. M. K. Bhamawat, "E-waste managementglobal scenario: a review," Journal of Environmental Research and Development, vol. 2, no. 4, pp. 817–823, 2008.
2. Organization for Economic Cooperation and Development, Extended Producer Responsibility: A Guidance Manual for Governments, Organization for Economic Cooperation and Development, Paris, France, 2001.
3. R. Widmer, H. Oswald-Krapf, D. Sinha-Khetriwal, M. Schnellmann, and H. Böni, "Global perspective on e-waste," Environmental Impact Assessment Review, vol. 25, pp. 436–458, 2005.

4. G. Gaidajis, K. Angelakoglou, and D. Aktsoglou, "E-waste: environmental problems and current management," Journal of Engineering Science and Technology Review, vol. 3, no. 1, pp. 193–199, 2010.
5. Ministry of Environment & Forest, Guidelines for Environmentally Sound Management of E-Waste, Ministry of Environment & Forest, New Delhi, India, 2008.
6. J. E. Hoffmann, "Recovering precious metals from electronic scrap," JOM, vol. 44, no. 7, pp. 43–48, 1992.
7. B. H. Robinson, "E-waste: an assessment of global production and environmental impacts," Science of the Total Environment, vol. 408, pp. 183–191, 2009.
8. K. Betts, "Producing usable materials from e-waste," Environmental Science Technology, vol. 42, pp. 6782–6783, 2008.
9. B. K. Gullett, W. P. Linak, A. Touati, S. J. Wasson, S. Gatica, and C. J. King, "Characterization of air emissions and residual ash from open burning of electronic wastes during simulate rudimentary recycling operations," Journal of Material Cycles and Waste Management, vol. 9, pp. 69–79, 2007.
10. C. A. Ibrahim, "E-waste management in Malaysia: current status. Presentation Slides for Bengkel Kajian Penyediaan Garis Panduan Mengenai Kriteria "Environmentally Sound Management" Ke Atas Kemudahan Pemerolehan Kembali E-Waste Di Malaysia," Palm Garden Hotel, Putrajaya, November 2013.
11. Organization for Economic Cooperation and Development, Guidance Manual on Environmentally Sound Management of Waste, Organization for Economic Cooperation and Development, Paris, France, 2007.
12. Ministry of Environment, Study on Criteria and Requirement on Environmentally Sound Management of Hazardous Wastes and Other Wastes, Ministry of Environment, Tokyo, Japan, 2011.

13. D. Sinha-Khetriwal, P. Kraeuchi, and M. Schwaninger, "A comparison of electronic waste recycling in Switzerland and in India," Environmental Impact Assessment Review, vol. 25, no. 5, pp. 492–504, 2005.
14. P. Rathore, S. Kota, and A. Chakrabarti, "Sustainability through remanufacturing in India: a case study on mobile handsets," Journal of Cleaner Production, vol. 19, no. 15, pp. 1709–1722, 2011.
15. U. Kumar and D. N. Singh, "E-waste management through regulations," vol. 3, no. 2, pp. 6–14, 2013.
16. A. Paratiamby and D. Victor, "Policy trends of e-waste management in Asia," Journal of Material Cycles and Waste Management, vol. 15, no. 4, pp. 411–419, 2013.
17. Organisation for Economic Co-operation and Development, Extended Producer Responsibility Phase 2 Case Studies on the German Packaging Ordinance, Organisation for Economic Co-operation and Development, Paris, France, 1998,https://www.olis.oecd.org/olis/1997doc.nsf/LinkTo/NT0000103A/$FILE/00044014.PDF.

Chapter 3

Energy and Environmental Performance of Bioethanol from Different Lignocelluloses

Lin Luo, Ester van der Voet, and Gjalt Huppes

Institute of Environmental Sciences (CML), Leiden University, P.O. Box 9518, 2300 RA Leiden, The Netherlands

ABSTRACT

Climate change and the wish to reduce the dependence on oil are the incentives for the development of alternative energy sources. The use of lignocellulosic biomass together with cellulosic processing technology provides opportunities to produce fuel ethanol with less competition with food and nature. Many studies on energy analysis and life cycle assessment of second-generation bioethanol have been conducted. However, due to the different methodology used and different system boundary definition, it is difficult to compare their results. To permit a direct comparison of fuel ethanol from different lignocelluloses in terms of energy use and environmental impact, seven studies conducted in our group were summarized in this paper, where the same technologies were used to convert biomass to ethanol, the same system boundaries were defined, and the same allocation procedures were followed. A complete set of environmental impacts ranging from global warming potential to toxicity aspects is used. The results provide an overview on the energy efficiency and environmental performance of using fuel ethanol derived from different feedstocks in comparison with gasoline.

INTRODUCTION

Climate change and the wish to reduce the dependence on oil are the incentives for the development of alternative energy sources. In view of the carbon dioxide reduction target agreed upon in the Kyoto protocol, a shift from fossil fuels to renewable resources is ongoing, also to secure the long-term energy supply at both national and international level. The European Commission demonstrated in 2007 that a 20% target for the overall share of energy from renewable sources and a 10% target for energy from renewable sources in transport would be appropriate and achievable objectives [1], though both targets have become subject of dispute since then. In the near term, liquid biofuels will still largely contribute to the target in transport sector due to the limited available technologies for fuels from other renewable sources. Especially bioethanol with its biorenewable nature, optimized production technology, and potential of greenhouse gas (GHG) mitigation already proved itself as an attractive alternative fuel.

Most of the current practice only concerns first-generation ethanol from conventional crops like corn, wheat, sorghum, potato, sugarcane, sugar beet, and so forth. Criticism is expressed on the first-generation bioethanol with regard to land use requirement and competition with food and nature. These issues have become the driving forces for the technology innovation towards second-generation ethanol from lignocellulosic feedstocks, both as agricultural coproducts and wastes and dedicated crops. EC Directive 2009 addressed the importance of commercializing second-generation biofuels [2]. The United States, as one of the leading nations in promoting biofuels, proposed that cellulosic ethanol must achieve 44% of the total biofuel production by 2020 [3].

With the development of the cellulosic technology being able to convert agricultural coproducts of often lower value to ethanol, two major questions are raised when comparing second-generation bioethanol to fossil fuel and first-generation bioethanol. (1) Is second-generation ethanol a better option for energy conservation? (2) What are the environmental benefits of second-generation ethanol? In order to answer the first question efforts were exerted on energy analysis of ethanol from corn stover [4, 5], switchgrass [6–9], and woodchip [7, 10]. Most of these studies yield a positive net energy value (NEV), which indicates that lignocelluloses are more favourable feedstocks than dedicated crops, that is, corn grain. However, Pimentel and Patzek [7] exceptionally reported negative NEV from switchgrass- and wood-ethanol systems concluding that cellulosic ethanol processes are more energy intensive than ethanol from corn grain. With respect to the second question many life cycle assessment (LCA) studies were conducted on ethanol from lignocelluloses such as corn stover [11–15], Switchgrass [9, 11], Miscanthus and willow [16], sugarcane bagasse [17, 18], cereal straw [19], woodchip and wood wastes [20–22], flax shives [23], and hemp hurds [24]. All these studies, to different extent, show environmental benefits especially in terms of reduced fossil resource depletion and GHG emissions.

However, these studies on both energy analysis and LCA of second-generation bioethanol raise a number of questions. First of all, there is insufficient consistency regarding the definition of system boundaries, with different choices made without explicit arguments. For instance, an ethanol refining system does not include the environmental impact from the production of cellulase enzyme for degrading cellulosic feedstocks [25], leaving out a major energy requiring process. Secondly, in different studies allocation methods are different and not systematic and mostly are unclearly stated. A high sensitivity to the allocation method has been reported for LCA results when evaluating carbon intensity and fossil energy consumption for bioethanol pathways [12, 20, 26, and 27]. The differences and ambiguities in the definition of system boundaries and allocation methodology made most of the studies incomparable.

To allow for a direct and meaningful comparison of different analysis, Farrell et al. [6] aligned methods and assumptions for six selected studies [7, 8, 28–30] and removed differences in the underlying data. Their focus, however, is mostly on corn-based ethanol. They indicate that calculations of NEV are highly sensitive to assumptions about both system boundaries and key parameter values and, as to content. They conclude that large-scale use of fuel ethanol certainly requires more sustainable practices in agriculture and advanced technologies, shifting from corn based to cellulosic ethanol production.

In order to have an overview of the energy intensity and environmental performance of bioethanol production from different lignocellulosic feedstocks, we have conducted five LCA studies at CML [12, 18, 23, 24, 31] and two studies on energy analysis [5, 32] of bioethanol from lignocelluloses, in all of which the same technologies were used to convert cellulose and hemicellulose to ethanol and to generate heat and power from lignin and wastes, the same system boundaries were defined, and the same allocation procedures were followed. In all LCA studies a comprehensive set of environmental impacts is used, ranging from abiotic depletion and GHG emissions to acidification and toxicity aspects. The present study summarizes and compares the results obtained from these five LCA studies and the two energy analyses, also in comparison with other studies.

LIFE CYCLE ASSESSMENT (LCA)

Methodology

All five case studies compared in the present study focus on LCA of fuel ethanol from lignocellulosic feedstocks, with application of different allocation methods and/or scenario analysis. In all studies advanced cellulosic processing technology is assumed to convert biomass to ethanol, the system boundaries are identical, and the same allocation procedure is applied. Therefore they are well comparable. The feedstocks in these studies are summarized as follows:

- Corn stover [12],
- Sugarcane and bagasse [18],
- Switchgrass [31],
- Flax shives [23],
- Hemp hurds [24].

In this section the methodologies used in all five studies are described. It is worth noting that study (2) presents a comparative LCA of ethanol production from solely sugar (current practice) and sugar plus bagasse (future case). As combining sugar and bagasse as feedstock for ethanol production is proposed to be a common practice in Brazil in the future, we did not perform an LCA on ethanol derived solely from bagasse. Thus the future case presented in study (2) is used to compare with other four studies, even though the feedstock includes not only the lignocellulose (bagasse) but also sugar.

Functional Unit and Alternatives

The functional unit in all studies is defined as power to wheels for one kilometre driving of a midsize flexible fuelled vehicle (FFV), which means that engine modification can be left out of account. Four fuel alternatives are considered: (1) conventional gasoline,

(2) a blend of 90% gasoline with 10% ethanol by volume (termed E10), (3) a blend of 15% gasoline with 85% ethanol by volume (termed E85), and (4) pure ethanol. In practice pure ethanol is not used as transport fuel; hence it is only a hypothetical case here for easy comparison with gasoline.

System Boundaries

The systems defined in all five studies are based on the "cradle-to-grave" approach, which includes crude oil extraction and refinery, with all transport required, and agriculture production of lignocellulosic biomass, harvest and transport of biomass to ethanol refinery, production of ethanol and its coproducts, blend and transport all fuels to regional storages, and the final use phase of fuels in vehicle driving. The production and transport of chemicals used in all processes are also taken into account. Moreover, emissions from capital goods production and waste management are included. However, emissions and wastes associated with the production and disposal of the FFV are outside the system boundaries.

In all five studies different coproducts are produced from either agricultural production or ethanol refinery, as listed in Table 1. In fact, in study (1), (2), (4), and (5) electricity is also cogenerated from the ethanol refinery; however, it is fully utilized by the refinery and no surplus can be considered as co-product. In study (2) extra electricity needs to be purchased from local grid due to the insufficiency of the cogenerated electricity inside the refinery.

Table 1: Product and coproducts from agriculture and ethanol refinery

Study	Products			
	Agriculture		Ethanol refinery	
	Product	Coproducts	Product	Coproducts
(1) Luo et al. [12]	Corn	Stover	Ethanol	—
(2) Luo et al. [18]	Sugarcane	—	Ethanol	Sugar

(3) Bai et al. [31]	Switchgrass	—	Ethanol	Electricity
(4) González García et al. [23]	Shives	Fibres & linseed	Ethanol	—
(5) González García et al. [24]	Hurds	Fibres & dust	Ethanol	—

Data Sources and Software

Life cycle inventory (LCI) data used in all studies are obtained from different sources, as summarized in Table2. The completeness of data may differ between sources; therefore, the Ecoinvent database from Swiss Centre of Life Cycle Inventories [33] is used as a standard whenever possible, as this source has long learning experience and involves a very broad range of processes (around 4,000). Software package Chain Management by Life Cycle Assessment (CMLCA) is used for the analysis [34].

Table 2: Data sources for the life cycle inventories of the five studies

Subsystem	Data source
Corn agriculture	U.S. Life-Cycle Inventory Database [35]
Sugarcane agriculture	Macedo et al. [36]
Switchgrass agriculture	Bullard and Metcalfe [37]; Nemecek and Kägi [38]
Flax agriculture	Nemecek et al. [39], Hauschild [40], Audsley et al. [41], Arrouays et al. [42], EMEP/CORINAIR [43]
Hemp agriculture	Nemecek et al. [39], Hauschild [40], Audsley et al. [41], Arrouays et al. [42], EMEP/CORINAIR [43]
Biomass transport	Spielmann et al. [44]
Stover-ethanol production	Aden et al. [45]
Sugarcane-ethanol production	Efe et al. [46]
Switchgrass-ethanol production	Guerra Miguez et al. [47]

Flax shives-ethanol production	Aden et al. [45]
Hemp hurds-ethanol production	Aden et al. [45]
Ethanol transport	Spielmann et al. [44]
Gasoline production & transport	Swiss Centre of Life Cycle Inventories [33]
Emissions from capital goods production	EIPRO Database [48]
Emissions from vehicle driving	Kelly et al. [49], Reading et al. [50]
Background processes	Swiss Centre of Life Cycle Inventories [33]

Allocation Method

The allocation procedure in multiproduct processes has been always one of the most critical issues in LCA. The ISO 14040 and 14044 [51, 52] recommend a stepwise procedure for allocation. First of all, allocation should be avoided whenever possible through subdivision of certain processes or by expanding system boundaries so as to include the additional functions related to them. If allocation cannot be avoided, methods reflecting underlying physical relationships shall be applied, of how process inputs and outputs change due to a quantitative change in products and functions delivered. This is often filled in as allocation based on mass or energy content of the coproducts. To the extent that physical relations cannot be established, other relevant variables like economic values of the coproducts can be used to allocate. This last option is similar to the cost allocation methods used in managerial accounting [53, 54].

In all five studies, one allocation method is applied, followed by a sensitivity analysis when two or more allocation methods can be applied. In some studies methods based on mass or energy content cannot be applied systematically. For instance, in the case of electricity as a co-product in ethanol refinery, mass allocation is not applicable; in the case of sugar as a co-product, energy content allocation cannot be used. Therefore, in the present study, in order to permit a fair comparison, economic allocation, which is used in all five studies, is applied. In no case the more general ISO requirements on physical allocation have been used or met. The multiproduct processes in these studies are given as follows:

- Corn agriculture, where corn and stover are produced;
- Ethanol production from sugarcane and bagasse, where ethanol and sugar are produced;
- Ethanol production from switchgrass, where ethanol and electricity are produced;
- Flax agriculture, where flax fibres, shives, and linseed are produced;
- Hemp agriculture, where hemp fibres, hurds, and dust are produced.

For the gasoline production, the allocations were taken as currently implemented in the Ecoinvent database by its designers, using different methods for different parts of the product system. This shortcoming is less relevant when comparing different biofuel options. The partitioning factors-based economic values in the five studies are given in Table 3. In study (4) and (5) the authors also conducted sensitivity analysis on the price of shives and hurds. For the comparison in the present study, an average price of flax shives and the price of hemp hurds in the base case scenario are taken.

Table 3: Partitioning factors for economic allocation in all five studies

Study	Multiproducts	Partitioning factor
(1) Luo et al. [12]	Stover/corn	0.118/0.882
(2) Luo et al. [18]	Ethanol/sugar	0.837/0.163
(3) Bai et al. [31]	Ethanol/electricity	0.894/0.106
(4) González García et al. [23]	Shives/fibers/linseed	0.042/0.911/0.047
(5) González García et al. [24]	Hurds/fibers/dust	0.122/0.863/0.015

Impact Assessment

The categories selected for impact assessment in all five studies are as follows:

- Abiotic depletion potential (ADP),
- Global warming potential (GWP),
- Ozone layer depletion potential (ODP),
- Photochemical oxidation potential (POCP),
- Human and ecotoxicity potential (HTP & ETP),
- Acidification Potential (AP),
- Eutrophication Potential (EP).

The environmental impacts addressed here reflect the differences between operations of vehicles fuelled with gasoline, E10, E85, and pure ethanol. The results are normalized to the "world total" in order to compare the importance of each impact. Overall evaluation requires weighting, which has not been applied in this study. The Handbook on Life Cycle Assessment [55] states: "Weighting is an optional step for all non-comparative assertions; there is no best available method and there is no recommended set of weighting factors." Nevertheless, for actual decision making some sort of weighting always is required.

Results Comparison and Discussion

The LCA results of fuel ethanol from the five different feedstocks compared with gasoline for each impact category are presented in Figure 1. In order to compare the relative size of results in different impact categories, they are normalized to "world total 1995" and are all set to the same scale. This normalised score indicates the percentage the system would contribute to the world total score on that impact, for the reference year taken.

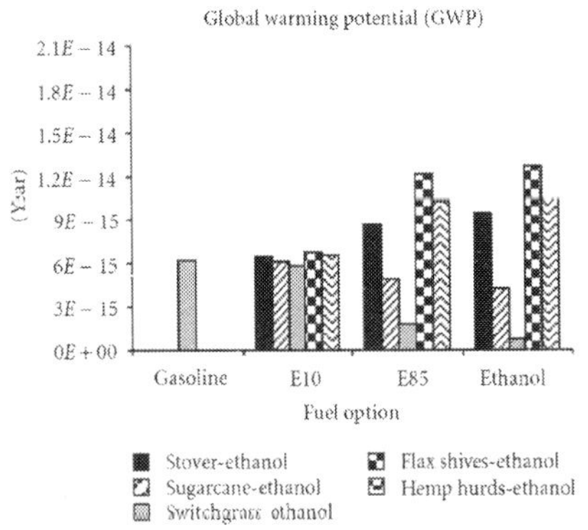
Global warming potential (GWP)
(Year)
2.1E − 14
1.8E − 14
1.5E − 14
1.2E − 14
9E − 15
6E − 15
3E − 15
0E + 00
Gasoline
E10
E85
Ethanol
Fuel option
Stover-ethanol
Sugarcane-ethanol
Switchgrass ethanol
Flax shives-ethanol
Hemp hurds-ethanol

(a)

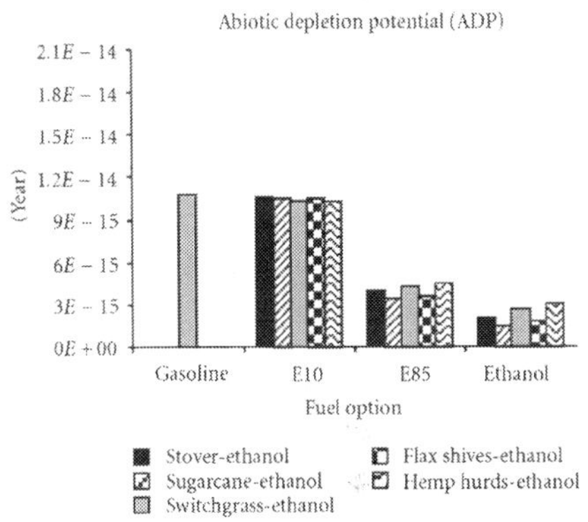
Abiotic depletion potential (ADP)
(Year)
2.1E − 14
1.8E − 14
1.5E − 14
1.2E − 14
9E − 15
6E − 15
3E − 15
0E + 00
Gasoline
E10
E85
Ethanol
Fuel option
Stover-ethanol
Sugarcane-ethanol
Switchgrass-ethanol
Flax shives-ethanol
Hemp hurds-ethanol

(b)

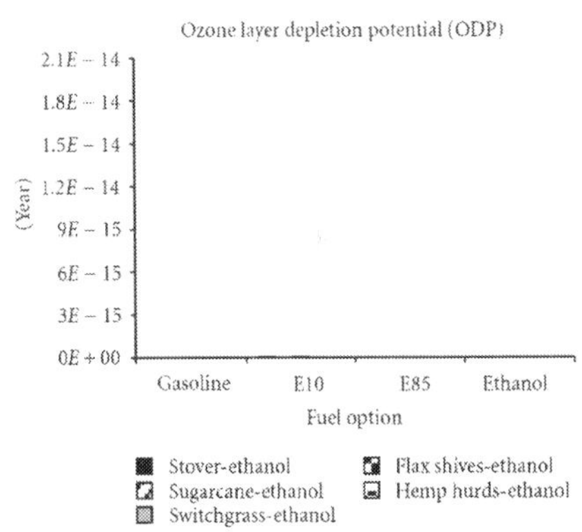
Ozone layer depletion potential (ODP)
(Year)
2.1E − 14
1.8E − 14
1.5E − 14
1.2E − 14
9E − 15
6E − 15
3E − 15
0E + 00
Gasoline
E10
E85
Ethanol
Fuel option
Stover-ethanol
Sugarcane-ethanol
Switchgrass-ethanol
Flax shives-ethanol
Hemp hurds-ethanol

(c)

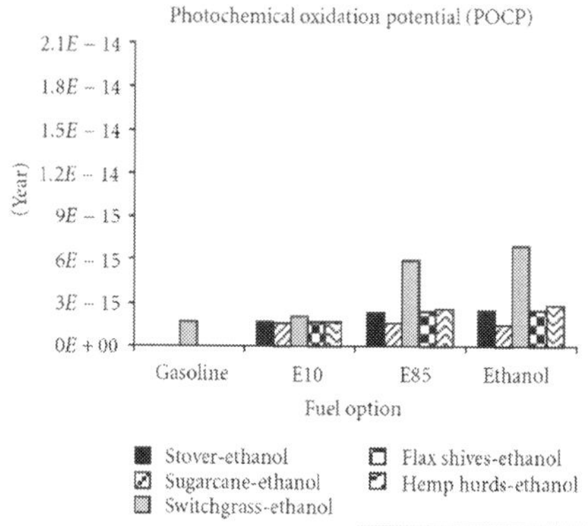
Photochemical oxidation potential (POCP)
(Year)
2.1E − 14
1.8E − 14
1.5E − 14
1.2E − 14
9E − 15
6E − 15
3E − 15
0E + 00
Gasoline
E10
E85
Ethanol
Fuel option
Stover-ethanol
Sugarcane-ethanol
Switchgrass-ethanol
Flax shives-ethanol
Hemp hurds-ethanol

(d)

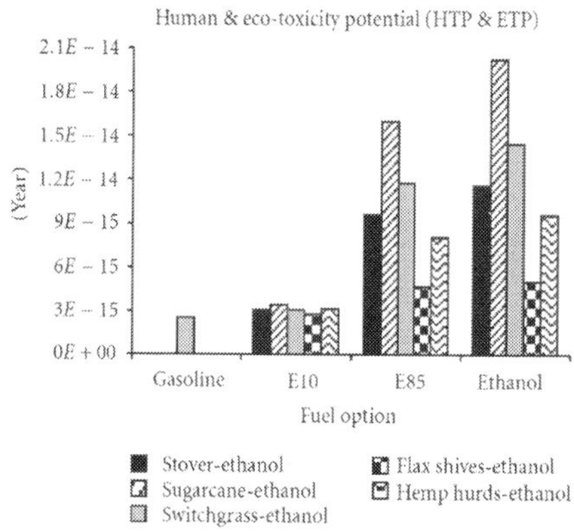
Human & eco-toxicity potential (HTP & ETP)
(Year)
2.1E − 14
1.8E − 14
1.5E − 14
1.2E − 14
9E − 15
6E − 15
3E − 15
0E + 00
Gasoline
E10
E85
Ethanol
Fuel option
Stover-ethanol
Sugarcane-ethanol
Switchgrass-ethanol
Flax shives-ethanol
Hemp hurds-ethanol

(e)

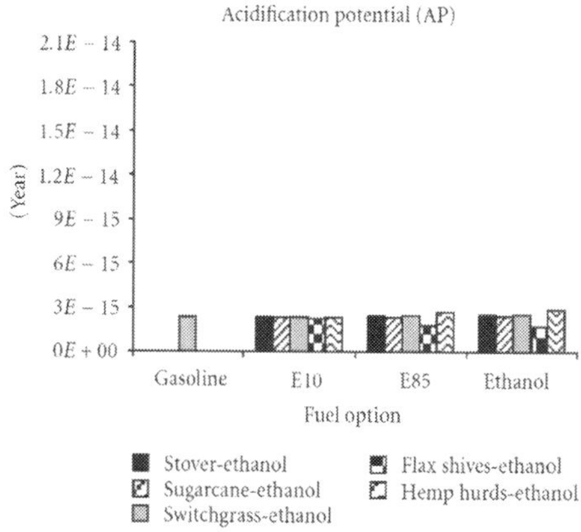
Acidification potential (AP)
(Year)
2.1E − 14
1.8E − 14
1.5E − 14
1.2E − 14
9E − 15
6E − 15
3E − 15
0E + 00
Gasoline
E10
E85
Ethanol
Fuel option
Stover-ethanol
Sugarcane-ethanol
Switchgrass-ethanol
Flax shives-ethanol
Hemp hurds-ethanol

(f)

[a]The unit of energy use is MJ per litre of ethanol produced from biorefinery. ormalized energy value for ethanol based on lower heating value (LHV).

Table 6: Net energy summary including coproducts, no allocation

Feedstock		Energy use (MJ/L[a])							
		Agriculture	Biorefinery	Total input	Ethanol[b]	Coproducts in agriculture[c]	Coproducts in biorefinery[d]	Total output	Net energy value (NEV)
Stover	Luo et al. [5]	10.0	0.5	10.5	21.2	85.2	0.2	106.6	96.1
Corn	Patzek [56]	9.9	17.0	26.9	21.2	25.9	4.1	51.2	24.3
	Pimentel and Patzek [7]	10.0	17.0	27.0	21.2	25.9	1.9	49.0	22.0
	Shapouri and McAloon [30]	5.3	15.2	20.5	21.2	24.0	7.3	52.5	32.0
	Graboski [29]	5.6	16.6	22.2	21.2	24.6	4.1	49.9	27.7
	Diaz de Oliveira et al. [28]	6.3	14.1	20.4	21.2	24.6	4.1	49.9	29.5
	Wu et al. [9]	6.6	12.5	19.1	21.2	24.6	4.0	49.8	30.7
Switchgrass	Wu et al. [9]	2.4	1.0	3.4	21.2	0.0	4.8	26.0	22.6
Sugarcane	Hadzhiyska et al. (base case) [32]	6.5	0.1	6.6	21.2	0.0	32.7	53.9	47.3
	Hadzhiyska et al. (future case) [32]	2.7	0.5	3.2	21.2	0.0	12.6	33.8	30.6
	Pimentel and Patzek [57]	2.6	3.4	6.0	21.2	0.0	0.0	21.2	15.2
	Diaz de Oliveira et al. [28]	5.6	0.6	6.2	21.2	0.0	1.1	22.3	16.1
	Macedo et al. [36]	2.0	0.3	2.3	21.2	0.0	3.0	24.2	21.9

[a]The unit of energy use is MJ per litre of ethanol produced from biorefinery.

[b]Normalized energy value for ethanol based on lower heating value (LHV).

[c]In stover-ethanol study it refers to the corn as a coproduct in the agriculture; in the other six studies on corn-ethanol it refers to the harvested stover (60%, dry mass basis) as a coproduct, and the values are estimated in this study.

[d]In stover-ethanol and switchgrass-ethanol study it refers to electricity; in corn-ethanol studies it refers to a range of products from corn milling, such as dried distiller grains, corn gluten feed, and corn oil; in sugarcane-ethanol study it refers to electricity surplus or/and bagasse.

When coproducts are not included meaning that all the inputs and outputs are 100% allocated to stover, three studies on corn-ethanol result in negative net energy values due to their then seemingly more intensive agriculture practice. The other three studies on corn ethanol yield very small positive NEV values compared to the ones of ethanol from sugarcane and cellulosic feedstocks (stover and switchgrass). However, stover is a relatively bad feedstock due to its intensive agriculture production process. The reason why the energy use in agriculture per litre of ethanol in stover-ethanol case is the highest is that the yield of ethanol from stover is lower than the one from corn. In order to produce 1 litre of ethanol more stover is needed—the average ethanol yield of ethanol from corn is 0.4l/kg, while from stover it is only 0.3l/kg.

When the energy credits (taken as the energy content of the product) are taken into account for all coproducts, all studies yield positive net energy values. The highest NEV of stover-ethanol attributes to the co-product in agriculture—corn. The resulting NEVs from corn-ethanol studies become comparable with those from ethanol derived from switchgrass and sugarcane. In the study conducted by Hadzhiyska et al. [32] sugar is coproduced in the ethanol refinery in both base and future case; the values of net energy are significantly higher than the ones from other studies without sugar coproduction. This indicates that outcomes of net energy calculations depend on whether or not taking the energy values of coproducts into account.

As mentioned earlier, in our sugarcane-ethanol study, the base case demonstrates the current practice in Brazilian ethanol

32. D. Hadzhiyska, L. Luo, and E. van der Voet, "An energy analysis of ethanol from sugar cane," CML Report, Leiden, The Netherlands, 2009.
33. "Swiss Centre of Life Cycle Inventories," June 2007, http://www.ecoinvent.org/.
34. "Chain Management by Life Cycle Assessment (CMLCA)," May 2007,http://cml.leiden.edu/software/software-cmlca.html.
35. "U.S. Life-Cycle Inventory Database," June 2007, http://www.nrel.gov/lci/.
36. I. C. Macedo, M. R. L. V. Leal, and J. E. A. R. da Silva, "Assessment of greenhouse gas emissions in the production and use of fuel ethanol in Brazil," Report, Government of the State of São Paulo and Secretariat of the Environment, São Paulo, Brazil, 2004.
37. M. Bullard and P. Metcalfe, "Estimating the energy requirement and CO2 emissions from production of the perennial grasses miscanthus, switchgrass and reed canary grass," Tech. Rep. ETSU B/U1/00645/REP.
38. T. Nemecek and T. Kägi, "Life cycle inventories of Swiss and European agricultural production systems," Ecoinvent Report 15, Swiss Centre for Life Cycle Inventories, Dübendorf, Switzerland, 2007.
39. T. Nemecek, A. Heil, O. Huguenin, et al., "Life cycle inventories of agricultural production systems,"Ecoinvent Report 15, Swiss Centre for Life Cycle Inventories, Dübendorf, Switzerland, 2004.
40. M. Z. Hauschild, "Estimating pesticide emissions for LCA of agricultural products," in Agricultural Data for Life Cycle Assessments, B. P. Weidema and M. J. G. Meeusen, Eds., vol. 2, pp. 64–79, LCANet Food, The Hague, The Netherlands, 2000.
41. E. Audsley, S. Alber, R. Clift, et al., "Harmonization of environmental life cycle assessment for agriculture,"

Concerted Action AIR3-CT94-2028, European Commission. DG VI Agriculture. SRI, Silsoe, UK, 1997.

42. D. Arrouays, J. Balesdent, J. C. Germon, P. A. Jayet, J.F. Soussana, and P. Stengel, "Contribution à la lutte contre l›effet de serre. Stocker du carbone dans les sols agricoles de France? Expertise scientifique collective," Rapport d›expertise réalisé par INRA à la demande du Ministère de l›Ecologie et du Développement Durable, INRA, Paris, France, 2002.
43. EMEP/CORINAIR, "Atmospheric emission inventory guidebook," Tech. Rep. 11, European Environment Agency, Copenhagen, Denmark, 2006.
44. M. Spielmann, T. Kägi, P. Stadler, and O. Tietje, "Life cycle inventories of transport services," Ecoinvent Report 14, Swiss Centre for Life Cycle Inventories, Dübendorf, Switzerland, 2004.
45. A. Aden, M. Ruth, and K. Ibsen, "Lignocellulosic biomass to ethanol process design and economics utilizing co-current dilute acid prehydrolysis and enzymatic hydrolysis for corn stover," Tech. Rep. NREL/TP-510-32438, 2002.
46. Ç. Efe, A. J. J. Straathof, and L. A. M. van der Wielen, "Technical and economic feasibility of production of ethanol from sugarcane and sugarcane bagasse," in B-Basic Internal Report, 2005.
47. O. Guerra Miguez, D. Siva Siddarthan, and D. A. Suárez Zuluaga, "Technical and economic feasibility of production of ethanol from switchgrass," in B-Basic Internal Report, Delft University of Technology, Delft, The Netherlands, 2009.
48. A. Tukker, G. Huppes, J. B. Guinée, et al., "Environmental Impact of Products (EIPRO)—Analysis of the life cycle environmental impacts related to the final consumption of the EU-25," 2005.
49. K. J. Kelly, B. K. Bailey, T. C. Coburn, W. Clark, and P. Lissiuk, "Federal test procedure emissions test results from ethanol variable-fuel vehicle Chevrolet Luminas," in Society

for Automotive Engineers, International Spring Fuels and Lubricants Meeting, Dearborn, Mich, USA, 1996.

50. A. H. Reading, J. O. W. Norris, E. A. Feest, and E. L. Payne, "Ethanol emissions testing," AEAT unclassified E&E/DDSE/02/021, Issue 3, 2002.

51. ISO Norm 14040, "Life cycle assessment: principles and framework," Environmental management, 2006.

52. ISO Norm 14044, "Life cycle assessment: requirement and guidelines," Environmental management, 2006.

53. G. Huppes, Macro-Environmental Policy: Principles and Design, Elsevier, Amsterdam, the Netherlands, 1993.

54. J. B. Guinée, R. Heijungs, and G. Huppes, "Economic allocation: examples and derived decision tree,"International Journal of Life Cycle Assessment, vol. 9, no. 1, pp. 23–33, 2004.

55. J. B. Guinée, Ed., Handbook on Life Cycle Assessment—Operational Guide to the ISO Standard, Kluwer Academic Publishers, Dodrecht, the Netherlands, 2002.

56. T. W. Patzek, "Thermodynamics of the corn-ethanol biofuel cycle," Critical Reviews in Plant Sciences, vol. 23, no. 6, pp. 519–567, and 2004.

57. D. Pimentel and T. Patzek, "Ethanol production: energy and economic issues related to U.S. and Brazilian sugarcane," Natural Resources Research, vol. 16, no. 3, pp. 235–242, 2007.

58. A. R. Ravishankara, J. S. Daniel, and R. W. Portmann, "Nitrous oxide (N2O): the dominant ozone-depleting substance emitted in the 21st century," Science, vol. 326, no. 5949, pp. 123–125, 2009.

Chapter 4

A Review of Deterministic Optimization Methods in Engineering and Management

Ming-Hua Lin[1], Jung-Fa Tsai[2], and Chian-Son Yu[1]

[1]Department of Information Technology and Management, Shih Chien University, No. 70, Ta-Chih Street, Taipei City 10462, Taiwan

[2]Department of Business Management, National Taipei University of Technology, No. 1, Section 3, Chung-Hsiao E. Road, Taipei City 10608, Taiwan

ABSTRACT

With the increasing reliance on modeling optimization problems in practical applications, a number of theoretical and algorithmic contributions of optimization have been proposed. The approaches developed for treating optimization problems can be classified into deterministic and heuristic. This paper aims to introduce recent advances in deterministic methods for solving signomial programming problems and mixed-integer nonlinear programming problems. A number of important applications in engineering and management are also reviewed to reveal the usefulness of the optimization methods.

INTRODUCTION

The field of optimization has grown rapidly during the past few decades. Many new theoretical, algorithmic, and computational contributions of optimization have been proposed to solve various problems in engineering and management. Recent developments of optimization methods can be mainly divided into deterministic and heuristic approaches. Deterministic approaches take advantage of the analytical properties of the problem to generate a sequence of points that converge to a global optimal solution. Heuristic approaches have been found to be more flexible and efficient than deterministic approaches; however, the quality of the obtained solution cannot be guaranteed. Moreover, the probability of finding the global solution decreases when the problem size increases. Deterministic approaches (e.g., linear programming, nonlinear programming, and mixed-integer nonlinear programming, etc.) can provide general tools for solving optimization problems to obtain a global or an approximately global optimum. With the increasing reliance on modeling optimization problems in real applications, a number of deterministic methods for optimization problems have been presented. This paper focuses on discussing and reviewing the recent advances in deterministic optimization approaches.

Optimization methods have been applied in different fields, such as finance [1–3], allocation and location problems [4–6], engineering design [7–12], system and database design [13–17], chemical engineering design and control [18–22], and molecular biology [23]. For additional literature on real-world applications or developments of optimization methods, readers may refer to the following works. Mockus et al. [24] treated network problems, knapsack, travelling salesman, flow-shop problems, and batch process scheduling problems by Bayesian heuristic approaches. Grossmann [25] discussed the global optimization algorithms and their applications in chemical engineering design. Bomze et al. [26] investigated decision support systems and techniques for solving molecular structures, queuing systems, image reconstruction, location analysis, and process network synthesis problems. Migdalas et al. [27] presented multilevel optimization algorithms and their applications. Mistakidis and Stavroulakis [28] studied engineering applications of the finite element method. De Leone et al. [29] proposed various interesting applications of high-performance software for nonlinear optimization. Hendrix [30] utilized global optimization techniques on environmental management, geometric design, robust product design, and parameter estimation. Corliss and Kearfott [31] presented a rigorous global search method on industrial applications. Floudas and Pardalos [32] studied optimization approaches in the fields of computational chemistry and molecular biology. Laguna and González-Velarde [33] discussed advanced computing tools for tackling various challenging optimization problems. Papalambros and Wilde [34] investigated the principles and practice of optimal engineering design. Atkinson et al. [35] gave a detailed discussion on optimum experimental design. Edgar et al. [36] explored the optimization methods of chemical processes. Pardalos and Romeijn [37] provided a more complete and broad spectrum of approaches including deterministic and heuristic techniques for dealing with global optimization problems. Tawarmalani and Sahinidis [38] provided an insightful and comprehensive treatment of convexification and global optimization of continuous and mixed-integer nonlinear programs. Hadjisavvas et al. [39] investigated generalized convexity and generalized monotonicity and offered an advanced and broad overview of the state of the field. Moreover, Floudas et al. [40] presented an overview of the research progress in optimization during 1998–2003, including the deterministic global optimization advances in mixed-integer nonlinear programming and related applications. Pintér [41] illustrated the applicability of global optimization modeling techniques and solution strategies to real-world problems such as agroecosystem management, assembly line design, bioinformatics, biophysics, cellular mobile network design, chemical product design, composite structure design, controller design for induction motors, electrical engineering design, feeding strategies in animal husbandry, the inverse position problem in kinematics, laser design, radiotherapy planning, robot design, and satellite data analysis. Mishra and Giorgi [42] presented results on invex function, and their properties in smooth and nonsmooth cases, pseudolinearity, and eta-pseudolinearity. Mishra et al. [43] discussed the Kuhn-Tucker optimality, Karush-Kuhn-Tucker necessary and sufficient optimality conditions in presence of various types of generalized convexity assumptions. Floudas and Gounaris [44] also presented an overview of the research progress in deterministic global optimization during the last decade (1998–2008).

Figure 1 gives an overview of the problem types related to optimization problems. Each type of problems has received substantial attention from the practitioners and the researchers in the last few decades. In this paper, we investigate the advances in deterministic global optimization of nonconvex nonlinear programming (NLP) problems and nonconvex mixed-integer nonlinear programming (MINLP) problems. For NLP problems, we focus on signomial programming problems that are an important class of NLP problems and have played a crucial role in applications.

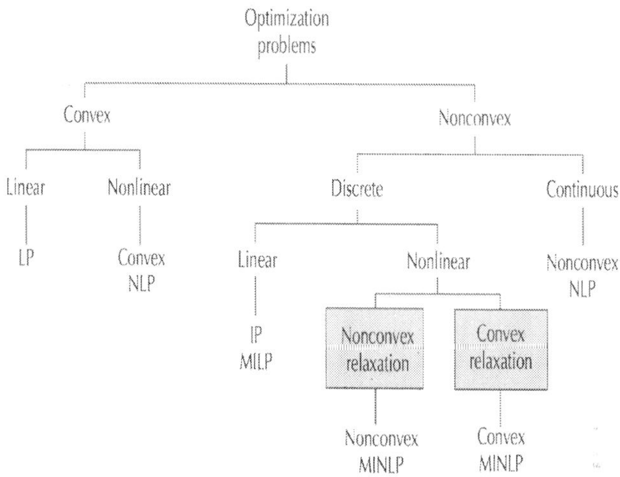

Figure 1: Overview of problem types related to optimization problems.

The rest of this paper is organized as follows. Section 2 discusses the deterministic methods for signomial programming problems. Section 3 reviews the theoretical and algorithmic developments of mixed-integer nonlinear programming problems. Conclusions are made in Section 4.

SIGNOMIAL PROGRAMMING

Signomial programming (SP) is an optimization technique for solving a class of nonconvex nonlinear programming problems. Although SP problems occur frequently in engineering and management science, SP problems with nonconvex functions are still difficult to be solved to obtain a global optimum. The term geometric program (GP) was introduced by Duffin et al. [7] since the analysis of geometric programs relied heavily upon geometric-arithmetic mean inequality. The early work by Duffin and Peterson [45] solved the posynomial geometric program analytically via the dual problem. Then, Duffin [46] developed a numerical method for GPs based on solving a sequence of linear programs [47]. For solving SP problems, Duffin and Peterson [48] reformulated an SP problem as a geometric program with reversed constraints. The reversed constraints give rise to a nonconvex feasible region that the local minima for SP problems are not guaranteed to be global minima [49]. The developed methods for SP can be divided into two approaches. The first class of SP approaches includes various heuristic techniques such as genetic algorithms, simulated annealing, tabu search, ant colony optimization, and particle swarm optimization. Although the heuristic methods have the advantage of easy implementation and offer a better potential for complex problems, the obtained solution is not guaranteed to be a globally optimal solution. The second class of SP approaches is the deterministic method. For example, Maranas and Floudas [50], Floudas and Pardalos [51], Maranas and Floudas [22], and Floudas [19–21] developed global optimization algorithms for solving SP problems based on the exponential variable transformation, the convex relaxation, and the branch and bound type algorithm. These methods transform the original nonconvex problem into a convex problem and then solve it to obtain the global optimum. The use of the logarithmic/exponential transformation in global optimization algorithms on SP problems restricts these exponential-based methods to handle the problems with strictly positive variables. Although positive variables are employed frequently to represent engineering and scientific systems, it is also common to introduce nonpositive variables in modeling the management problems or the system behavior, such as investment decisions, stresses, temperatures, electrical currents, velocities, and accelerations. For treating free variable x, Pörn et al. [52] suggested a simple translation, $x+\tau=e^x$. However, inserting the transformed result into the original signomial term will bring additional signomial terms and therefore increase the computation burden. Tsai et al. [11] proposed an approach to treat zero boundary signomial discrete programming problems and suggested some convexification rules. Li and Tsai [53], Tsai and Lin [54–56], Tsai et al. [57], Tsai [58], and Li and Lu [59] applied convexification strategies and piecewise linearization techniques to solve SP problems with free discrete/continuous variables. However, the optimal solution obtained is an approximate solution by the piecewise linearization approach. Lin and Tsai [60] presented a generalized method to solve signomial discrete programming problems with free variables for finding exactly alternative optima. Tsai and Lin [61] also integrated the convexification techniques and the bounding schemes to solve a posynomial geometric program with separable functions for finding a global optimal solution efficiently.

Convexification strategies for signomial terms are important techniques in global optimization for SP problems. With different convexification approaches, an SP problem can be reformulated into another convex program solvable to obtain an approximately global optimum. For solving SP problems, Pörn et al. [52] integrated the exponential transformation and piecewise linear approximations for reformulating nonconvex signomial problems. The results were extended by Björk [62], Björk et al. [63], and Pörn et al. [64], by including certain power transformations for convexification of nonconvex signomial terms. They discussed that the right choice of transformation for convexifying nonconvex signomial terms has a clear influence on the efficiency of the optimization approach. The concept of power convex functions is introduced to improve the solution efficiency for certain SP problems. T. Westerlund and J. Westerlund [65] proposed the generalized geometric programming extended cutting plane (GGPECP) algorithm for nonconvex optimization problems by using the cutting plane and transformation techniques. The GGPECP algorithm was described in more detail in Westerlund [66]. Lundell and Westerlund [67] and Lundell et al. [68] combined the GGPECP algorithm with an optimization framework for the transformations used to convexify the signomial terms into a signomial global optimization algorithm. The signomial global optimization algorithm was further extended by Lundell and Westerlund [69, 70]. Lin and Tsai [71] and Tsai and Lin [56] also presented similar reformulation and range reduction techniques to enhance the computational efficiency for solving SP problems.

For solving an SP problem, the above-mentioned convexification techniques are used to reformulate the original SP problem into a convex and underestimating problem solvable by a standard mixed-integer nonlinear programming (MINLP) solver [72–81]. Different transformations for positive and negative signomial terms have been proposed and discussed by Björk et al. [63], Westerlund [66], Lundell and Westerlund [67], Pörn et al. [64], Lundell et al. [68], and Lundell and Westerlund [69, 70]. For a positive signomial term $cx_1^{\alpha_1} x_2^{\alpha_2} \ldots\ldots x_m^{\alpha_m} x_{m+1}^{\alpha_{m+1}} \ldots\ldots x_n^{\alpha_n} \left(c > 0, \alpha_1, \ldots, \alpha_m > 0 \text{ and } \alpha_{m+1}, \ldots, \alpha_n < 0\right)$, they suggested either the exponential transformation (ET) or the power convex transformation (PCT) is applied based on the characteristics of the problems. The ET strategy and the PCT strategy are described as follows [63].

The ET strategy:

$$cx_1^{\alpha_1} x_2^{\alpha_2} \cdots x_m^{\alpha_m} x_{m+1}^{\alpha_{m+1}} \cdots x_n^{\alpha_n} \Longleftrightarrow \begin{cases} x_i = e^{X_i}, & i = 1, \ldots, m, \\ c\dfrac{e^{\alpha_1 X_1 + \cdots + \alpha_m X_m}}{x_{m+1}^{|\alpha_{m+1}|} \cdots x_n^{|\alpha_n|}}. & \end{cases} \tag{2.1}$$

The PCT strategy: this technique aims at constructing 1-convex signomial terms. First, transform all variables with positive exponents by an inverse transformation (IT), x=X-1, except the one with the greatest exponent denoted as .αmax Let S be defined as $S = \sum_{i=1}^{n} |\alpha_i| - \alpha_{max}$. If $\alpha_{max} < S + 1$, then transform the variable with the exponent αmax to that with the exponent S+1. If $\alpha_{max} > S + 1$, then change one of the ITs, with the exponent to αj, where $x_j = X_j^{-\tau}$ so that $\alpha_{max} = S + 1 + (\tau - 1)\alpha_j$.

Since some negative signomial terms may exist in SP problems, they suggested the potential transformation (PT) for a negative signomial term $cx_1^{\alpha_1} x_2^{\alpha_2} \ldots\ldots x_m^{\alpha_m} x_{m+1}^{\alpha_{m-1}} \ldots\ldots x_n^{\alpha_n} \left(c < 0, \alpha_1, \ldots, \alpha_m > 0 \text{ and } \alpha_{m+1}, \ldots, \alpha_n < 0\right)$ expressed as follows.

The PT strategy:

$$cx_1^{\alpha_1} x_2^{\alpha_2} \cdots x_m^{\alpha_m} x_{m+1}^{\alpha_{m+1}} \cdots x_n^{\alpha_n} \Longleftrightarrow \begin{cases} x_i = X_i^{1/R}, & i = 1, \ldots, m, \\ x_i = X^{-1/R}, & i = m+1, \ldots, n, \\ cX_1^{\alpha_1/R} \cdots X_m^{\alpha_m/R} X_{m+1}^{|\alpha_{m+1}|/R} \cdots X_n^{|\alpha_n|/R}, & R = \sum_{i=1}^{n} |\alpha_i|. \end{cases} \tag{2.2}$$

In addition to convexification strategies, convex envelopes and convex underestimators of nonconvex functions are frequently applied in global optimization algorithms such as the BB algorithm [19, 82, 83] to underestimate the nonconvex functions. A good convex underestimator should be as tight as possible and contain minimal number of new variables and constraints thus to improve the computational effect of processing a node in a branch-bound tree [40].

Tawarmalani and Sahinidis [84] developed the convex envelope and concave envelope for x/y over a unit hypercube, proposed a semidefinite relaxation of x/y, and suggested convex envelopes for functions of the form $f(x)y^2$ and f(x)/y. Ryoo and Sahinidis [85] studied the use of arithmetic intervals, recursive arithmetic intervals, logarithmic transformation, and exponential transformation for multilinear functions. Tawarmalani et al. [86] studied the role of disaggregation in leading to tighter linear programming relaxations. Tawarmalani and Sahinidis [38] introduced the convex extensions for lower semicontinuous functions, proposed a technique for constructing convex envelopes for nonlinear functions, and studied the maximum separation distance for functions such as x/y. Tawarmalani et al. [87] studied 0-1 hyperbolic programs and developed eight mixed-integer convex reformulations. Liberti and Pantelides [88] proposed a nonlinear continuous and differentiable convex envelope for monomials of odd degree, derived its linear relaxation, and compared to other relaxations. Björk et al. [63] studied convexifications for signomial terms, introduced properties of power convex functions, compared the effect of the convexification schemes for heat exchanger network problems, and studied quasiconvex convexifications. Meyer and Floudas [89] studied trilinear monomials with positive or negative domains, derived explicit expressions for the facets of the convex and concave envelopes, and showed that these outperform the previously proposed relaxations based on arithmetic intervals or recursive arithmetic intervals. Meyer and Floudas [90] presented explicit expressions for the facets of convex and concave envelopes of trilinear monomials with mixed-sign domains. Tardella [91] studied the class of functions whose convex envelope on a polyhedron coincides with the convex envelope based on the polyhedron vertices and proved important conditions for a vertex polyhedral convex envelope. Meyer and Floudas [92] described the structure of the polyhedral convex envelopes of edge-concave functions over polyhedral domains using geometric arguments and proposed an algorithm for computing the facets of the convex envelopes.

Caratzoulas and Floudas [93] proposed novel convex underestimators for trigonometric functions, which are trigonometric functions themselves. Akrotirianakis and Floudas [94, 95] introduced a new class of convex underestimators for twice continuously differentiable nonlinear programs, studied their theoretical properties, and proved that the resulting convex relaxation is improved compared to the αBB one. Meyer and Floudas [90] proposed two new classes of convex underestimators for general C^2 nonlinear programs, which combine the BB underestimators within a piecewise quadratic perturbation, derived properties for the smoothness of the convex underestimators, and showed the improvements over the classical αBB convex underestimators for box-constrained optimization problems.

Three popular convex underestimation methods, arithmetic intervals (AIs) [96], recursive arithmetic intervals (rAIs) [50, 85, 96], and explicit facets (EFs) for convex envelopes of trilinear monomials [89, 90], are effective to underestimate a trilinear term $x_1\ x_2\ x_3$ for x_i to be bounded variables. However, these existing methods have difficulty to treat a posynomial function. According to Ryoo and Sahinidis [85], for underestimating a multilinear function $x_1\ x_{2 \ldots\ldots}\ x_n$ with n variables, the AI scheme needs to use $\prod_{k=2}^{n-1}\Theta_k^{\binom{n}{k}}\sum_{i=1}^{\lfloor n/2\rfloor}\binom{n}{2i}$ linear constraints maximally. Θ_k denotes the number of linear functions that the AI scheme generates to lower bound k-cross-product terms,k=2,3,…,n-1 . Since the number of linear constraints of convex envelopes for a multilinear function with n variables grows doubly exponentially in n, AI bounding scheme may only treat $n \leq 3$ cases. It is more difficult for AI to treat a posynomial function for n>3 cases. Moreover, applying rAI scheme to underestimate a multilinear function $x_1x_2...x_n$ needs to use the maximum of exponentially many 2^{n-1} linear inequalities. Therefore, the rAI bounding scheme has difficulty to treat posynomial functions as well as the AI scheme. EF [89, 90] provided the explicit facets of the convex and concave envelopes of trilinear monomials and demonstrated that these result in tighter bounds than the AI and rAI techniques. An important difference between EF and other bounding schemes is that these explicit facets are linear cuts, which were proven to define the convex envelope. Explicit facets (EFs) of the convex envelope are effective in treating general trilinear monomials, but the derivation of explicit facets for the convex envelope of general multilinear monomials and signomials is an open problem. Li et al. [97] and Lu et al. [98] proposed a novel method for the convex relaxation of posynomial functions. The approach is different from the work of Maranas and Floudas [50], which provided an alternative way of generating convex underestimators for generalized geometric programming problems via the exponential transformation and linear underestimation of the concave terms. Applications of this approach include the area of process synthesis and design of separations, phase equilibrium, nonisothermal complex reactor networks, and molecular conformation problems (e.g., [99–101]).

MIXED-INTEGER NONLINEAR PROGRAMMING

Mixed-integer nonlinear programming (MINLP) problems involving both continuous and discrete variables arise in many applications of engineering design, chemical engineering, operations research, and management. Biegler and Grossmann [102] provided a retrospective on optimization techniques that have been applied in process systems engineering. They indicated that design and synthesis problems have been dominated by NLP and MINLP models. With the increasing reliance on modeling optimization problems in practical problems, a number of theoretical and algorithmic contributions of MINLP have been proposed. Many deterministic methods for convex MINLP problems have been reviewed by Biegler and Grossmann [102], Grossmann [103], and Grossmann and Biegler [104]. The methods include branch-and-bound (BB) [53, 72, 77], generalized benders decomposition (GBD) [76], outer-approximation (OA) [73, 74, 79], extended cutting plane method (ECP) [81], and generalized disjunctive programming (GDP) [78]. The BB method can find the global solution only when each subproblem can be solved to global optimality. The GBD method, the OA method, and the ECP method cannot solve optimization problems with nonconvex constraints or nonconvex objective functions because the subproblems may not have a unique optimum in the solution process. The GDP models address discrete/continuous optimization problems that involve disjunctions with nonlinear inequalities and logic propositions. The objective functions and the constraints in the GDP problem are assumed to be convex and bounded [56].

For deterministic optimization methods, these optimization problems are characterized by the convexity of the feasible domain or the objective function and may involve continuous and/or discrete variables. Although continuous and discrete optimization problems constitute two classes of global optimization problems, they primarily differ in the presence or absence of convexity rather than other features. Since the convexity of the objective function or the feasible domain is very important, understanding how to convexify the nonconvex parts is an essential area of research. As long as the formulated problem is a convex problem, efficient numerical methods are available to treat the optimization problem. However, optimization problems often include nonconvex functions that cannot be dealt with by the standard local optimization techniques to guarantee global optimality efficiently. For solving nonconvex or large-scale optimization problems, deterministic methods may not be easy to derive an optimal solution within reasonable time due to the high complexity of the problem.

Sherali et al. [105] presented an extension of the reformulation linearization technique (RLT) that is designed to exploit special structures and explored the strengthening of the RLT constraints through conditional logical expressions. Sinha et al. [106] studied a solvent design problem that is constructed as a nonconvex MINLP problem. They identified the sources of nonconvexities in the properties and solubility parameter design constraints and proposed linear underestimators based on a multilevel representation approach for the functions. A reduced space branch-and-bound global optimization algorithm was then presented for solving a single component blanket wash design problem. Pörn et al. [52] introduced different convexification strategies for nonconvex MINLP problems with both posynomial and negative binomial terms in the constraints. Harjunkoski et al. [107] studied the trim loss minimization problem for the paper converting industry and formulated the model as a nonconvex MINLP. They also proposed transformations for the bilinear terms based on linear representations and convex expressions. Pörn and Westerlund [108] proposed a cutting plane method for addressing global MINLP problems with pseudoconvex objective function and constraints and tested the proposed method on several benchmark problems arising in process synthesis and scheduling applications. Parthasarathy and El-Halwagi [109] studied a systematic framework for the optimal design of condensation, which is an important technology for volatile organic compounds, and formulated the problem as a nonconvex MINLP model. They also proposed an iterative global optimization approach based on physical insights and active constraint principles that allow for decomposition and efficient solution and applied it to a case study for the manufacture of adhesive tapes. Adjiman et al. [82, 83, 110, 111] proposed two global optimization approaches, SMIN-αBB and αGMIN-BB, for nonconvex MINLP problems based on the concept of branch-and-bound. These two approaches rely on optimization or interval-based variable-bound updates to enhance efficiency. Although one possible approach to circumvent nonconvexities in nonlinear optimization models is reformulation, for instance, using the exponential transformation to treat the generalized geometric programming problems in which a signomial term $x_1^\alpha x_2^\beta$ is transferred into an exponential term $e^{\alpha \ln x_1 + \beta \ln x_2}$, the exponential transformation technique can only be applied to strictly positive variables and is thus unable to deal with nonconvex problems with free variables. Tsai et al. [11] proposed an approach to treat zero boundary optimization problems and suggested some convexification rules for the signomial terms with only three nonnegative discrete/integer variables. Björk and Westerlund [112] studied the global optimization of heat exchanger network synthesis through the simplified superstructure representation that allows only series and parallel schemes and applied convexification approaches for signomials by piecewise linear approximations. They also formulated convex MINLP lower bounding models using the Patterson formula for the log mean temperature difference considering both isothermal and nonisothermal mixing. Ostrovsky et al. [113] studied nonconvex MINLP models in which most variables are in the nonconvex terms and the number of linear constraints is much larger than the number of nonlinear constraints for solvent design and recovery problems. The work presents a tailored branch-and-bound approach using linear underestimators for tree functions based on a multilevel function representation and shows that there is a significant reduction in the branching variable space. Tawarmalani and Sahinidis [114] developed a branch and bound framework for the global optimization of MINLP problems. The framework involves novel linear relaxation schemes, a Lagrangian/linear duality-based theory for domain and range reduction, and branching strategies that guarantee finiteness of the solution sequence for certain classes of problems. They also discuss implementation issues and present computational results with a variety of benchmark problems. Kesavan et al. [115] presented outer-approximation algorithms for finding an optimal solution of a separable nonconvex MINLP program. Emet and Westerlund [116] conducted a computational comparison of solving a cyclic chromatographic separation problem using MINLP methods and reported that the extended cutting plane method compares favourably against traditional outer-approximation and branch-and-bound methods. A review of the recent advances in MINLP optimization of planning and design problems in the process industry was presented by Kallrath [117]. Tawarmalani and Sahinidis [118] introduced a polyhedral branch-and-cut approach in global optimization. Their algorithm exploits convexity in order to generate the polyhedral cuts and relaxations for multivariate nonconvex problems. Meyer and Floudas [119] studied superstructures of pooling networks, which are important to the petrochemical, chemical, and wastewater treatment industries, and formulated this generalized pooling problem as a nonconvex MINLP problem that involves many bilinear terms in the constraint functions. They proposed a global optimization algorithm based on a novel piecewise linear reformulation-linearization technique (RLT) formulation. Karuppiah and Grossmann [120] addressed the problem of optimal synthesis of an integrated water system, where water using processes and water treatment operations are jointly considered. The designed MINLP model was solved with a new deterministic spatial branch and contract algorithm, in which piecewise under- and overestimators are used for constructing the relaxations at each node. Bergamini et al. [121] formulated an MINLP model for the global optimization of heat exchanger networks and presented a new solution methodology that is based on outer-approximation and utilizes piecewise underestimation. Rigorous constraints obtained from physical insights are also included in the formulation, and the authors reported computationally efficient global solutions for problems with up to nine process streams. Tsai and Lin [54, 56] proposed a method for solving a signomial MINLP problem with free variables by the convexification strategies and piecewise linearization techniques. However, the optimal solution obtained is an approximate solution by the piecewise linearization approach. Karuppiah et al. [122] presented an outer-approximation algorithm to globally solve a nonconvex MINLP formulation that corresponds to the continuous time scheduling of refinery crude oil operations. The solution procedure relies on effective mixed-integer linear relaxations that benefit from additional cuts derived after spatially decomposing the network. Foteinou et al. [123] presented a mixed-integer optimization framework for the synthesis and analysis of regulatory networks. Their approach integrates prior biological knowledge regarding interactions between genes and corresponding transcription factors, in an effort to minimize the complexity of the problem. Misener et al. [124] proposed an extended pooling problem to maximize the profit of blending reformulated gasoline on a predetermined network structure of feed stocks, intermediate storage tanks, and gasoline products subject to applicable environmental standards. They formulated the problem as a nonconvex MINLP model due to the presence of bilinear, polynomial, and fractional power terms. A mixed-integer linear programming relaxation of the extended pooling problem is proposed for several small- to large-scale test cases. Misener et al. [125] introduced a formulation for the piecewise linear relaxation of bilinear functions with a logarithmic number of binary variables and computationally compared their performance of this new formulation to the best performing piecewise relaxations with a linear number of binary variables. They also unified the new formulation into the computational tool APOGEE that globally optimizes standard, generalized, and extended pooling problems. Westerlund et al. [126] considered some special but fundamental issues related to convex relaxation techniques in nonconvex MINLP optimization, especially for optimization problems including nonconvex inequality constraints and their relaxations.

The alternative global optima of an MINLP problem can be found if more than one solution satisfies the same optimal value of the objective function. In practice, alternative optima are useful because they allow the decision maker to choose from many solutions without experiencing any deterioration in the objective function. For the case involving only 0-1 variables, Balas and Jeroslow [127] introduced the well-known binary cut with only one constraint and no additional variables. Duran and Grossmann [73] used this binary cut in their OA algorithm to exclude binary combinations. Tawarmalani and Sahinidis [38] mentioned that BARON can identify the K best solutions for a mixed-integer nonlinear program, where K is an option specified by the user. Tsai et al. [57] proposed a general integer cut to identify all alternative optimal solutions of a general integer linear programming problem.

Lin and Tsai [60] proposed a generalized method to find multiple optimal solutions of an MINLP problem with free variables by means of variable substitution and convexification strategies. The problem is first converted into another convex MINLP problem solvable to obtain an exactly global optimum. Then, a general cut is utilized to exclude the previous solution and an algorithm is developed to locate all alternative optimal solutions.

CONCLUSIONS

Given the rapid advances in computing technology over the past decades, large optimization theories and algorithms have been proposed to solve various real-world engineering and management problems. Therefore, to give a systematic overview of the extant literature is a challenge and motivates this study, particularly for that the field of optimization has grown and evolved rapidly. This work first reviewed methods for continuous variable optimization and survey advances in signomial programming. Then, mixed-integer nonlinear programming methods for optimization problems with discrete components were introduced. Contributions related to theoretical and algorithmic developments, formulations, and applications for these two classes of optimization problems were also discussed.

Although deterministic approaches take advantage of analytical properties of the problem to generate a sequence of points that converge to a global solution, heuristic approaches have been found to be more flexible and efficient than deterministic approaches. For solving nonconvex or large-scale optimization problems, deterministic methods may not be easy to derive a globally optimal solution within reasonable time due to the high complexity of the problem. Heuristic approaches therefore are presented to reduce the computational time of solving an optimization problem, but the obtained solution is not guaranteed to be a feasible or globally optimal solution. These two types of optimization methods have different pros and cons. Therefore, integrating deterministic and heuristic approaches may be a good way of solving large-scale optimization problems for finding a global optimum. It is hoped that this paper will stimulate further research on developing more advanced deterministic and heuristic methods to enhance the computational efficiency of finding a globally optimal solution for various real application problems.

ACKNOWLEDGMENTS

The research is supported by Taiwan NSC Grants NSC 99-2410-H-158-010-MY2 and NSC 99-2410-H-027-008-MY3.

REFERENCES

1. W. Sharpe, "A linear programming approximation for the general portfolio analysis,"Journal of Financial and Quantitative Analysis, vol. 6, pp. 1263–1275, 1971.
2. B. Stone, "A linear programming formulation of the general portfolio selection model,"Journal of Financial and Quantitative Analysis, vol. 8, pp. 621–636, 1973.
3. M. R. Young, "A minimax portfolio selection rule with linear programming solution,"Management Science, vol. 44, no. 5, pp. 673–683, 1998.
4. C. S. Chen, S. M. Lee, and Q. S. Shen, "An analytical model for the container loading problem," European Journal of Operational Research, vol. 80, no. 1, pp. 68–76, 1995. ·
5. L. Faina, "A global optimization algorithm for the three-dimensional packing problem,"European Journal of Operational Research, vol. 126, no. 2, pp. 340–354, 2000.
6. H. L. Li, C. T. Chang, and J. F. Tsai, "Approximately global optimization for assortment problems using piecewise linearization techniques," European Journal of Operational Research, vol. 140, no. 3, pp. 584–589, 2002. ·

7. R. J. Duffin, E. L. Peterson, and C. Zener, Geometric programming: Theory and application, John Wiley & Sons Inc., New York, 1967.
8. J. S. Arora, Introduction to Optimum Design, McGraw Hill, New York, NY, USA, 1989.
9. E. Sandgren, "Nonlinear integer and discrete programming in mechnical design optimization," Journal of Mechanisms, Transmissions, and Automation in Design, vol. 112, no. 2, pp. 223–229, 1990.
10. J. F. Fu, R. G. Fenton, and W. L. Cleghorn, "A mixed integer-discrete -continuous programming method and its application to engineering design optimization,"Engineering Optimization, vol. 17, pp. 263–280, 1991.
11. J. F. Tsai, H. L. Li, and N. Z. Hu, "Global optimization for signomial discrete programming problems in engineering design," Engineering Optimization, vol. 34, no. 6, pp. 613–622, 2002.
12. J. F. Tsai, "Global optimization for signomial discrete programming problems in engineering design," Engineering Optimization, vol. 42, no. 9, pp. 833–843, 2010.
13. D. Rotem, G. A. Schloss, and A. Segev, "Data allocation for multidisk databases," IEEE Transactions on Knowledge and Data Engineering, vol. 5, no. 5, pp. 882–887, 1993.
14. X. Lin and M. Orlowska, "An integer linear programming approach to data allocation with the minimum total communication cost in distributed database systems,"Information Sciences, vol. 85, no. 1–3, pp. 1–10, 1995.
15. R. Sarathy, B. Shetty, and A. Sen, "A constrained nonlinear 0-1 program for data allocation," European Journal of Operational Research, vol. 102, no. 3, pp. 626–647, 1997. ·
16. J. F. Tsai and H. L. Li, "Technical note—on optimization approach for multidisk vertical allocation problems," European Journal of Operational Research, vol. 165, no. 3, pp. 835–842, 2005.

17. M. H. Lin, "An optimal workload-based data allocation approach for multidisk databases," Data and Knowledge Engineering, vol. 68, no. 5, pp. 499–508, 2009.
18. H. S. Ryoo and N. V. Sahinidis, "Global optimization of nonconvex NLPs and MINLPs with applications in process design," Computers and Chemical Engineering, vol. 19, no. 5, pp. 551–566, 1995.
19. C. A. Floudas, "Global optimization in design and control of chemical process systems,"Journal of Process Control, vol. 10, no. 2, pp. 125–134, 2000.
20. C. A. Floudas, "Recent advances in global optimization for process synthesis, design and control: enclosure of all Solutions," Computers and Chemical Engineering, vol. 23, no. 1, pp. S963–S973, 1999.
21. C. A. Floudas, Deterministic Global Optimization: Theory, Methods and Application, Kluwer Academic Publishers, Boston, Mass, USA, 2000.
22. C. D. Maranas and C. A. Floudas, "Global optimization in generalized geometric programming," Computers and Chemical Engineering, vol. 21, no. 4, pp. 351–369, 1997. ·
23. J. G. Ecker, M. Kupferschmid, C. E. Lawrence, A. A. Reilly, and A. C. H. Scott, "An application of nonlinear optimization in molecular biology," European Journal of Operational Research, vol. 138, no. 2, pp. 452–458, 2002.
24. J. Mockus, W. Eddy, L. Mockus, and G. Reklaitis, Bayesian Heuristic Approach to Discrete and Global Optimization, Kluwer Academic Publishers, London, UK, 1996.
25. I.E. Grossmann, Global Optimization in Engineering Design, Kluwer Academic Publishers, London, UK, 1996.
26. I. M. Bomze, T. Csendes, R. Horst, and P. M. Pardalos, Developments in Global Optimization, Kluwer Academic Publishers, London, UK, 1997.
27. A. Migdalas, P. M. Pardalos, and P. Varbrand, Multilevel Optimization: Algorithms and Applications, Kluwer Academic Publishers, London, UK, 1997.

28. E. S. Mistakidis and G. E. Stavroulakis, Nonconvex Optimization in Mechanics: Algorithm, Heuristics and Engineering Applications by the F.E.M, Kluwer Academic Publishers, London, UK, 1997.
29. R. De Leone, A. Murli, P. M. Pardalos, and G. Toraldo, High Performance Software for Nonlinear Optimization, Kluwer Academic Publishers, London, UK, 1998.
30. E. M. T. Hendrix, Global optimization at work, Ph.D. thesis, Agriculture University Wageningen, Wageningen, The Netherlands, 1998.
31. G. F. Corliss and R. B. Kearfott, "Rigorous global search: industrial applications," inDevelopments in Reliable Computing, T. Csendes, Ed., pp. 1–16, Kluwer Academic Publishers, London, UK, 1999.
32. C. A. Floudas and P. M. Pardalos, Optimization in Computational Chemistry and Molecular Biology-Local and Global Approaches, Kluwer Academic Publishers, London, UK, 2000.
33. M. Laguna and J. L. González-Velarde, Computing Tools for Modeling Optimization and Simulation, Kluwer Academic Publishers, London, UK, 2000.
34. P. M. Papalambros and D. J. Wilde, Principles of Optimal Design-modeling and Computation, Cambridge University Press, Cambridge, UK, 2nd edition, 2000.
35. A. Atkinson, B. Bogacka, and A. A. Zhigljavsky, Optimum Design 2000, Kluwer Academic, London, UK, 2001.
36. T. F. Edgar, D. M. Himmelblau, and L. S. Lasdon, Optimization of Chemical Processes, McGraw-Hill, Boston, Mass, USA, 2nd edition, 2001.
37. P.M. Pardalos and H.E. Romeijn, Handbook of Global Optimization, vol. 2, Kluwer Academic Publishers, London, UK, 2002.
38. M. Tawarmalani and N. V. Sahinidis, "Convex extensions and envelopes of lower semi-continuous functions," Mathematical Programming A, vol. 93, no. 2, Ser. A, pp. 247–263, 2002.

39. N. Hadjisavvas, S. Komlósi, and S. S. Schaible, Handbook of Generalized Convexity and Generalized Monotonicity, Kluwer Academic Publishers, London, UK, 2005.
40. C. A. Floudas, I. G. Akrotirianakis, S. Caratzoulas, C. A. Meyer, and J. Kallrath, "Global optimization in the 21st century: advances and challenges," Computers and Chemical Engineering, vol. 29, pp. 1185–1202, 2005.
41. J. D. Pintér, Global Optimization: Scientific and Engineering Case Studies, Kluwer Academic Publishers, London, UK, 2006.
42. S. K. Mishra and G. Giorgi, Invexity and Optimization, Kluwer Academic Publishers, London, UK, 2008.
43. S. K. Mishra, S. Wang, and K. K. Lai, Generalized Convexity and Vector Optimization, Kluwer Academic Publishers, London, UK, 2009.
44. C. A. Floudas and C. E. Gounaris, "A review of recent advances in global optimization,"Journal of Global Optimization, vol. 45, no. 1, pp. 3–38, 2009.
45. R. J. Duffin and E. L. Peterson, "Duality theory for geometric programming," SIAM Journal on Applied Mathematics, vol. 14, pp. 1307–1349, 1966.
46. R. J. Duffin, "Linearizing geometric programs," SIAM Review, vol. 12, pp. 211–227, 1970.
47. S. Boyd, S.-J. Kim, L. Vandenberghe, and A. Hassibi, "A tutorial on geometric programming," Optimization and Engineering, vol. 8, no. 1, pp. 67–127, 2007.
48. R. J. Duffin and E. L. Peterson, "Geometric programming with signomials," Journal of Optimization Theory and Applications, vol. 11, pp. 3–35, 1973.
49. J. G. Ecker, "Geometric programming: methods, computations and applications," SIAM Review, vol. 22, no. 3, pp. 338–362, 1980. ·
50. C. D. Maranas and C. A. Floudas, "Finding all solutions of nonlinearly constrained systems of equations," Journal of Global Optimization, vol. 7, no. 2, pp. 143–182, 1995. ·

89. C. A. Meyer and C. A. Floudas, "Trilinear Monomials with Positive or Negative Domains: Facets of Convex and Concave Envelopes," in Frontiers in global optimization, C. A. Floudas and P. M. Pardalos, Eds., pp. 327–352, Kluwer Academic Publishers, Santorini, Greece, 2003.
90. C. A. Meyer and C. A. Floudas, "Trilinear monomials with mixed sign domains: facets of the convex and concave envelopes," Journal of Global Optimization, vol. 29, no. 2, pp. 125–155, 2004.
91. F. Tardella, "On the existence of polyhedral convex envelopes," in Frontiers in Global Optimization, C. A. Floudas and P. M. Pardalos, Eds., pp. 563–573, Kluwer Academic Publishers, Santorini, Greece, 2003.
92. C. A. Meyer and C. A. Floudas, "Convex envelopes for edge-concave functions,"Mathematical Programming B, vol. 103, no. 2, pp. 207–224, 2005.
93. S. Caratzoulas and C. A. Floudas, "Trigonometric convex underestimator for the base functions in Fourier space," Journal of Optimization Theory and Applications, vol. 124, no. 2, pp. 339–362, 2005.
94. I. G. Akrotirianakis and C. A. Floudas, "Computational experience with a new class of convex underestimators: box-constrained NLP problems," Journal of Global Optimization, vol. 29, no. 3, pp. 249–264, 2004.
95. I. G. Akrotirianakis and C. A. Floudas, "A new class of improved convex underestimators for twice continuously differentiable constrained NLPs," Journal of Global Optimization, vol. 30, no. 4, pp. 367–390, 2004.
96. A. S. E. Hamed, Calculation of bounds on variables and underestimating convex functions for nonconvex functions, Ph.D. thesis, The George Washington University, 1991.
97. H. L. Li, J. F. Tsai, and C. A. Floudas, "Convex underestimation for posynomial functions of positive variables," Optimization Letters, vol. 2, no. 3, pp. 333–340, 2008.

98. H. C. Lu, H. L. Li, C. E. Gounaris, and C. A. Floudas, "Convex relaxation for solving posynomial programs," Journal of Global Optimization, vol. 46, no. 1, pp. 147–154, 2010. ·
99. A. Aggarwal and C. A. Floudas, "Synthesis of general distillation sequences-nonsharp separations," Computers and Chemical Engineering, vol. 14, no. 6, pp. 631–653, 1990. ·
100. A. R. Ciric and C. A. Floudas, "A retrofit approach for heat exchanger networks,"Computers and Chemical Engineering, vol. 13, no. 6, pp. 703–715, 1989.
101. A. C. Kokossis and C. A. Floudas, "Optimization of complex reactor networks-II. Nonisothermal operation," Chemical Engineering Science, vol. 49, no. 7, pp. 1037–1051, 1994.
102. L. T. Biegler and I. E. Grossmann, "Retrospective on optimization," Computers and Chemical Engineering, vol. 28, no. 8, pp. 1169–1192, 2004.
103. I. E. Grossmann, "Review of nonlinear mixed-integer and disjunctive programming techniques," Optimization and Engineering, vol. 3, no. 3, pp. 227–252, 2002.
104. I. E. Grossmann and L. T. Biegler, "Part II. Future perspective on optimization,"Computers and Chemical Engineering, vol. 28, no. 8, pp. 1193–1218, 2004.
105. H. D. Sherali, W. P. Adams, and P. J. Driscoll, "Exploiting special structures in constructing a hierarchy of relaxations for mixed integer problems," Operations Research, vol. 46, no. 3, pp. 396–405, 1998. ·
106. M. Sinha, L. E. K. Achenie, and G. M. Ostrovsky, "Environmentally benign solvent design by global optimization," Computers and Chemical Engineering, vol. 23, no. 10, pp. 1381–1394, 1999.
107. I. Harjunkoski, T. Westerlund, and R. Pörn, "Numerical and environmental considerations on a complex industrial mixed integer non-linear programming (MINLP) problem," Computers and Chemical Engineering, vol. 23, no. 10, pp. 1545–1561, 1999.

108. R. Pörn and T. Westerlund, "A cutting plane method for minimizing pseudo-convex functions in the mixed integer case," Computers and Chemical Engineering, vol. 24, no. 12, pp. 2655–2665, 2000.
109. G. Parthasarathy and M. M. El-Halwagi, "Optimum mass integration strategies for condensation and allocation of multicomponent VOCs," Chemical Engineering Science, vol. 55, no. 5, pp. 881–895, 2000.
110. C. S. Adjiman, I. P. Androulakis, and C. A. Floudas, "Global optimization of MINLP problems in process synthesis and design," Computers and Chemical Engineering, vol. 21, no. 1, pp. S445–S450, 1997.
111. C. S. Adjiman, I. P. Androulakis, and C. A. Floudas, "Global optimization of mixed-integer nonlinear problems," AIChE Journal, vol. 46, no. 9, pp. 1769–1797, 2000.
112. K. M. Björk and T. Westerlund, "Global optimization of heat exchanger network synthesis problems with and without the isothermal mixing assumption," Computers and Chemical Engineering, vol. 26, no. 11, pp. 1581–1593, 2002.
113. G. M. Ostrovsky, L. E. K. Achenie, and M. Sinha, "On the solution of mixed-integer nonlinear programming models for computer aided molecular design," Computers and Chemistry, vol. 26, no. 6, pp. 645–660, 2002. ·
114. M. Tawarmalani and N. V. Sahinidis, "Global optimization of mixed-integer nonlinear programs: a theoretical and computational study," Mathematical Programming. A, vol. 99, no. 3, pp. 563–591, 2004.
115. P. Kesavan, R. J. Allgor, E. P. Gatzke, and P. I. Barton, "Outer approximation algorithms for separable nonconvex mixed-integer nonlinear programs," Mathematical Programming A, vol. 100, no. 3, pp. 517–535, 2004.
116. S. Emet and T. Westerlund, "Comparisons of solving a chromatographic separation problem using MINLP methods," Computers and Chemical Engineering, vol. 28, no. 5, pp. 673–682, 2004.

117. J. Kallrath, "Solving planning and design problems in the process industry using mixed integer and global optimization," Annals of Operations Research, vol. 140, pp. 339–373, 2005.
118. M. Tawarmalani and N. V. Sahinidis, "A polyhedral branch-and-cut approach to global optimization," Mathematical Programming. B, vol. 103, no. 2, pp. 225–249, 2005.
119. C. A. Meyer and C. A. Floudas, "Global optimization of a combinatorially complex generalized pooling problem," AIChE Journal, vol. 52, no. 3, pp. 1027–1037, 2006.
120. R. Karuppiah and I. E. Grossmann, "Global optimization for the synthesis of integrated water systems in chemical processes," Computers and Chemical Engineering, vol. 30, no. 4, pp. 650–673, 2006.
121. M. L. Bergamini, N. J. Scenna, and P. A. Aguirre, "Global optimal structures of heat exchanger networks by piecewise relaxation," Industrial and Engineering Chemistry Research, vol. 46, no. 6, pp. 1752–1763, 2007.
122. R. Karuppiah, K. C. Furman, and I. E. Grossmann, "Global optimization for scheduling refinery crude oil operations," Computers and Chemical Engineering, vol. 32, no. 11, pp. 2745–2766, 2008.
123. P. T. Foteinou, E. Yang, G. K. Saharidis, M. G. Ierapetritou, and I. P. Androulakis, "A mixed-integer optimization framework for the synthesis and analysis of regulatory networks," Journal of Global Optimization, vol. 43, no. 2-3, pp. 263–276, 2009.
124. R. Misener, C. E. Gounaris, and C. A. Floudas, "Mathematical modeling and global optimization of large-scale extended pooling problems with the (EPA) complex emissions constraints," Computers and Chemical Engineering, vol. 34, no. 9, pp. 1432–1456, 2010.
125. R. Misener, J. P. Thompson, and C. A. Floudas, "Apogee: global optimization of standard, generalized, and extended pooling problems via linear and logarithmic partitioning schemes," Computers and Chemical Engineering, vol. 35, no. 5, pp. 876–892, 2011.

126. T. Westerlund, A. Lundelf, and J. Westerlund, "On convex relaxations in nonconvex optimization," Chemical Engineering Transactions, vol. 24, pp. 331–336, 2011.

127. E. Balas and R. Jeroslow, "Canonical cuts on the unit hypercube," SIAM Journal on Applied Mathematics, vol. 23, pp. 61–69, 1972.

Chapter 5

Rethinking Petroleum Products Certification

Thiago Feital[1], Públio Lima[2], José Carlos Pinto[1], Maurício B. de Souza Jr[3], Gilberto Xavier[4, 5], Mario Jorge Lima[4], and Marcel Joly[4, 6]

[1]Programa de Engenharia Química/COPPE, Universidade Federal do Rio de Janeiro, 68502 Rio de Janeiro, RJ, and Brazil

[2]Escola Politécnica, Universidade Federal do Rio de Janeiro, 68529 Rio de Janeiro, RJ, and Brazil

[3]Departamento de Engenharia Química/EQ, Universidade Federal do Rio de Janeiro, 68542 Rio de Janeiro, RJ, and Brazil

[4]PETROBRAS Petróleo Brasileiro S.A., Rio de Janeiro, RJ, Brazil

[5]Faculdade de Tecnologia de Resende, Universidade do Estado do Rio de Janeiro, Rodovia Presidente Dutra Km 298, Resende, RJ, Brazil

[6]Departamento de Engenharia Química, Universidade de São Paulo, São Paulo, SP, Brazil

ABSTRACT

Facing various challenges in the everchanging refining landscape, it is essential that refiners raise their operations to new levels of performance. Advances in in-line blending (ILB) technology accuracy and reliability have encouraged refiners to take a step forward. Having ILB as a precursor, a new methodology is in concern: the so-called in-line certification (ILC) procedure. Blending processes make use of in-line measurements which, at least in principle, can be used to certificate the product, if the precision and accuracy of available in-line measurements are comparable to measurements provided by standard off-line tests. Such procedure may allow for significant reduction in refinery's tank farming and product inventory, increase of process flexibility, and reliability with benefits to company image. The main limitations for real-world ILC applications in the oil industry remain at the legal and technological levels. This paper proposes novel concepts and foundations of a basic in-line certification model for petroleum products regarding current interdisciplinary challenges and promising solutions.

INTRODUCTION

Petroleum refining is one of the most important industries, comprising many different and complicated processes with various possible configurations. Globally, it processes more materials than any other industry [1]. Due to the scale and significance of this industry, it becomes more crucial to address the considerable challenges that the industry faces today and in the future.

Production planning and scheduling optimization are essential tasks to maximize refinery's profit margins and to remain in the competitive market. In this sense, several opportunities strictly related to refinery production optimization could be identified. At the strategic level, the supply chain optimization may be posed as a master problem, since there are numerous trade-offs between decisions made at the various nodes of its superstructure in which the refinery planning is embedded. At the tactical level, two major challenges should be addressed by refiners [2].

The first one is that refining industry needs to effectively evaluate process performances to identify options for producing desirable products and meeting increasing constrained environment regulations. The second challenge refers to major transformation from an industry of mainly producing fuels for transportation to one that makes a wide set of value-added products, including chemicals, speciality products, electricity, and hydrogen.

In a scenario in which the world economic growth slowed down and demand for petroleum products flattened or even decreased, it becomes clear that flexibility is the keyword for everybody involved in the oil sector [3]. In this sense, online remote characterization and real-time optimization of petroleum products emerge as instrumental technologies for competitive refineries [4]. Such technologies may have a special impact on refining profitability with environmental benefits when considered to provide chemical information of great importance for real-time adjustment of short-term scheduling decisions concerned with product formulation and refinery dispatch logistics.

Blending consolidates as the refinery's last chance to impact profitability at low investment levels and approximately 50% of about 700 refineries worldwide have implemented such technology [5]. Besides positively impacting the overall refinery production scheduling, in-line blending (ILB) operations at the end of the refinery allow for substantial benefits including end product giveaway minimization, lower inventory levels, optimized logistics and, therefore, reduced utility consummations and atmospheric emissions, economic savings in new hardware investment and maintenance, and operational safety.

Apart from typical economic benefits ranging from US$ 0.15 to 0.30 per crude oil processed [6], ILB represents the precursor technology for allowing an additional efficiency jump in refinery scheduling optimization: the so-called in-line certification (ILC) (Figure 1).

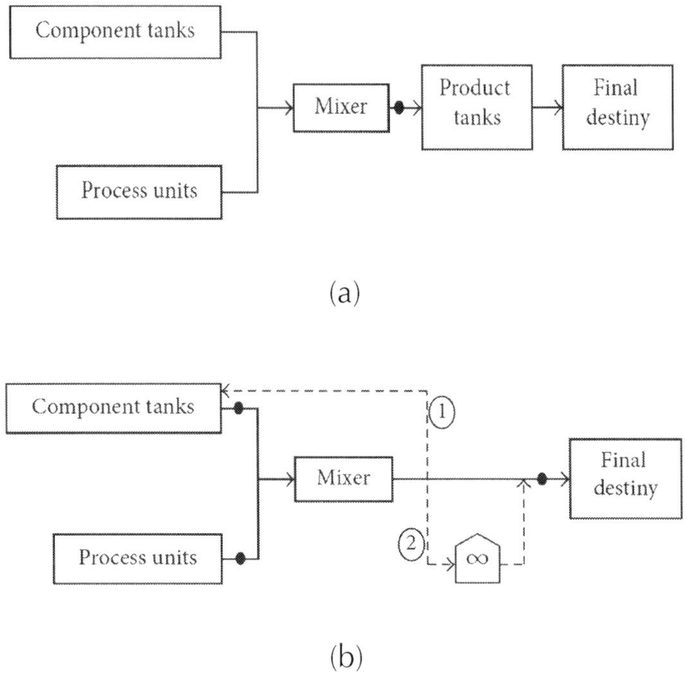

Figure 1: (a) Typical configuration of ILB units; (b) a virtual basic scheme for an ILC unit using two possible routes for off-spec fuels: (1) reblending or (2) infinite dilution. In both schemes, dashed circles represent on-line analyzers.

As the fundamental premise, an ILC-based approach should consider product dispatch to a third party before sending the official (i.e., legal) product quality certificate to it. ILC procedures are primarily intended to minimize refinery scheduling bottlenecks. However, it is not completely clear yet how large the benefits provided by ILC procedures can be. Since products can be delivered directly from processing to clients, some potential advantages that can probably be related to the implementation of ILC schemes are as follows.(i) Suppression of conventional operations, as tank homogenization, sampling, and laboratory testing, and this way saves manpower and shortening the refining production cycle time (Figure 2).(ii)Dead inventory minimization.(iii)Tankage minimization and associated maintenance costs optimization.(iv)Improvement of the process reliability and optimization (as process control techniques must necessarily be implemented in-line and in real time).(v)Enhanced flexibility to tightly specify different product grades according to real time customer needs.(vi)Shipping demurrage minimization. (vii)Enhanced energetic and environmental performances of the refinery storage area provided by an optimized scheduling.(viii)Staff safety and productivity increase provided by automation replacing manual operations.

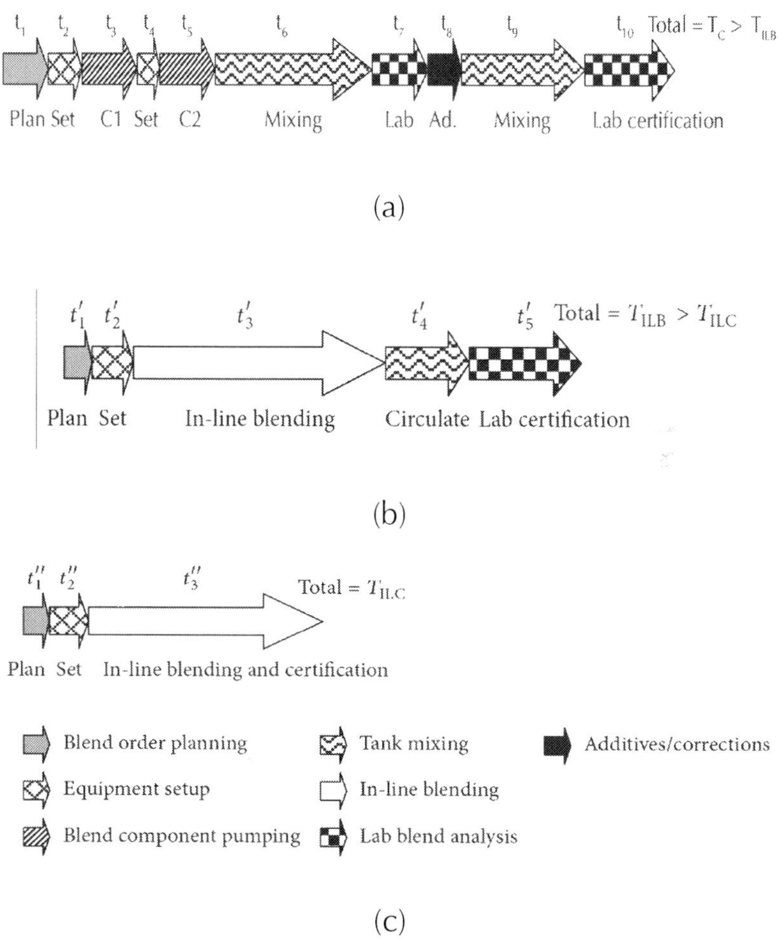

Figure 2: Time saving with ILC [7]. (a) Conventional time (T_C) operation based on manual tank blending and certification; (b) in-line blending

time (T_{ILB}) and conventional tank certification; (c) in-line certification time (T_{ILC}) of the finished blend product.

Such benefits must compensate the investment on technology and process upgrade required by ILC procedures, if actual implementation at plant site is sought. Apart from legal issues, the refinery's logistic environment will determine how expressive the opportunity for implementing an ILC-based operational philosophy is.

This paper is structured as follows. Section 2 proposes basic concepts and discusses an ILC-based business model. Some practical ILC implementation issues, with identification of current technical limitations, promising solutions, and required investments, are also discussed here. Section 3 discusses how the ILC scheme should work out in practice in order to be accepted by regulatory agencies and customers in general whereas Section 4 outlines some identified prerequisites to a generic ILC implementation. Section 5 presents the conclusions of the paper.

AN ILC-BASED CUSTODY TRANSFER MODEL

Custody transfer in the oil industry refers to the transactions involving transporting physical substance from one operator to another. Since presupposing product releasing before testing to a customer, an ILC-based approach implies new technical-commercial paradigms.

Model Definition

The first step required to support a new proposal or procedure is the definition of important related terms and the development of the basic ideas behind it. In-line certification is defined here as a computer-controlled procedure based on real-time data from on-line analyzers able to automatically measure stream qualities and perform actions over manipulated process variables to ensure

In-line certification should not be confounded with in-line delivery, which is exclusively based on the conventional certification of the products after delivery, usually performed to save time and storage at the plant. In-line certification, on the other hand, seeks for a definitive certification solution, as finished products are transferred directly to a third part as a fiscal batch and available in-line instrumentation and techniques are supposed to provide enough information about the product quality in reasonable time (real time) and with high confidence.

Currently, when not legally prohibited, both procedures are typically conditioned by regulatory agencies and are periodically subject to ad-hoc independent audit programs concerned with the computer-controlled system operation [5]. The goal of the audit program is to insure that fiscal batches data and information reported to the governmental entity are accurate and valid. Despite the lack of explicit conditions, regulatory agencies usually reserve their right to make inspections to assure compliance with the requirements (EPA 40CFR80.4).

In legal terms, the US legislation is virtually the most comprehensive one. Technical and legal procedure steps are described by US EPA in USA Code of Federal Regulation, 40CFR80, part 65, which allows refiners legally to blend directly to a third party before finishing product testing and CoA delivery. In Europe, shipping before testing practices have also been mentioned [8] but seem to be mediated by bilateral agreement between refiners and third parties. However, in both US and European cases, the number of refineries operating ILBC-like custody transfer is very limited and the third party is usually, if not always, a transporter company (pipeline or vessel), not a final client (distributor company). Conversely, from the aforementioned cases, in Brazil, the Petroleum National Agency, a governmental entity that regulates the Brazilian oil sector, currently does not allow for finished product shipping before the CoA emission.

The ILBC Hardware

An ILBC implementation presupposes process hardware properly designed with excellent instrumentation. Blend headers, static mixers, lines, pumps, and valves design should consider operational feasibility and safety accessories during transitory, steady-state, and fall-back conditions. Coriolis flowmeters have been extensively recommended for in-line blending systems [5, 8] and may help to persuade regulatory agencies to approve novel philosophies for custody transfer procedures, as ILBC. In order to reduce stream quality variability from (running) tank components, a tank mixer should be provided with the objective of allowing for inventory homogenization (during) before the fiscal batch production (route 2 in Figure 3). Whenever applicable, filters and coalescer devices for solid and free water removal should also be considered as basic blending system accessory.

At the design level, analytical redundancy should be a project premise. Since real data conciliation may be a valuable tool to the system reliability and early warning of problems, on-line analyzers at each component should be provided in order to produce redundant analytical information whenever possible. Temperature-controlled sampling systems should consider sampling points not susceptible to contaminants and located at critical process points, as components lines, blended fuel line as well as at the refinery boundary if the in-line certified product is then stocked in a tank or passes through it before shipping. Certified reference materials and prototype fuels (or protofuels), whose properties are known, require dedicated installations. Fast loops are required to allow for sample movement from process lines to the analyzer house (shelter) and vice versa (Figure 5).

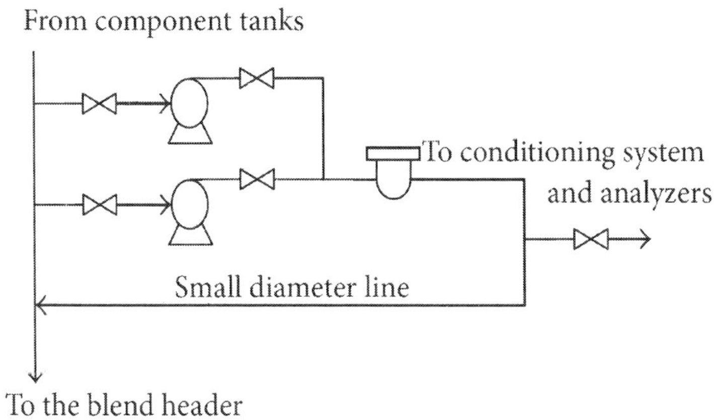

(a)

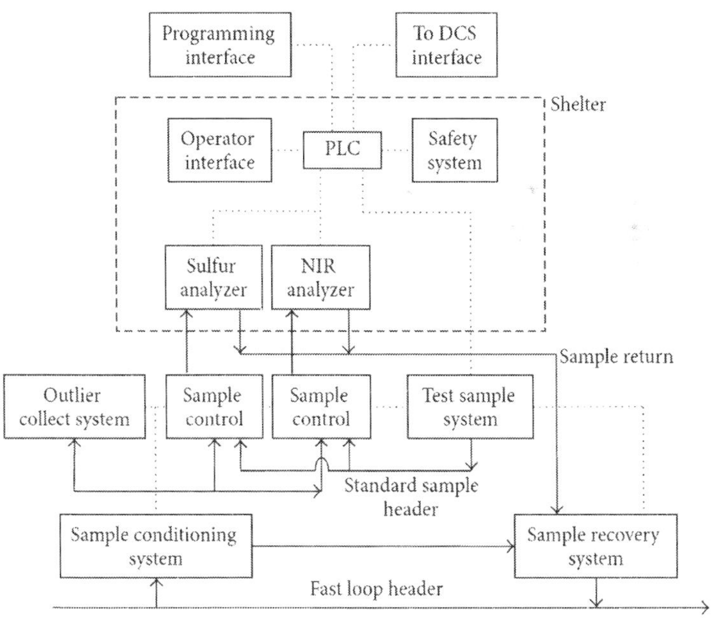

(b)

Figure 5: A typical fast loop scheme for refinery ILB. (a) Pumping and filtering; (b) the sample conditioning system.

The ILBC Software

As a real-time optimization business layer, the ILBC operation must agree with the refinery's objectives that are defined at the strategic and tactical levels. This is achievable through capturing the business knowledge in advanced decision-making tools, which should allow for increasing visibility into plant operations and communicating economic drivers to operations allowing them to identify and eliminate constraints by better understanding the planning economic objectives [12].

A schematic representation for ILBC software applications is proposed in Figure 6. In this figure, blending scheduling refers to an off-line optimization layer in which the refinery tactical planning is detailed into short-term operational decisions. Here, fiscal batches are defined and sequenced to execution throughout a time horizon aiming at satisfying the foreseen product requests by customers subject to refinery's operational, logistic, and economic constraints. The Blend Property Control (BPC) comprises at least two real-time applications: (a) the multivariable predictive control (MPC), which is responsible by the real-time blended product quality control, and (b) the on-line optimization process model, which should maximize or minimize some performance criteria subject to real-time constraints regarding the whole fiscal batch production. On the other hand, the Regulatory Blend Control (RBC) is responsible for manipulating component flows to match an optimal blend recipe defined at the BPC layer. For each predefined fiscal batch volume, the BPC application functions should include field equipment selection and lineup, pump sequencing control, appropriate ramps for blending start-up and conclusion, and additive injections. In addition, the RBC should also be responsible for implementing a trim fall-back strategy in case of system failure. The Key Performance Indicator (KPI) is a monitoring module from which the fiscal batch data is collected to compose the ECoA right after the batch conclusion.

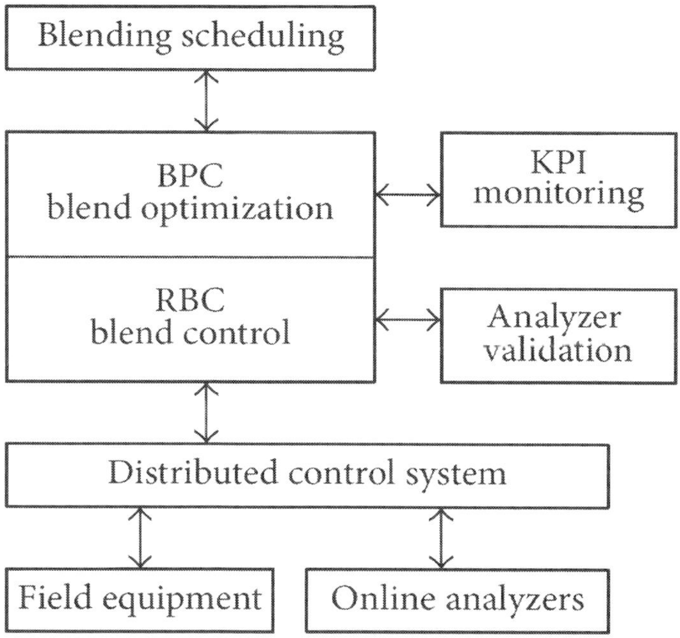

Figure 6: The ILBC software applications.

It seems clear that the implementation of advanced control schemes must certainly be considered during development of ILBC procedures, as process disturbances are inherent to process operation and lead to some degree of unavoidable process variability. Therefore, process variability must be rigorously considered during the implementation of the ILBC scheme, with the help of sound statistical analyses of measured data performed in-line and on real time.

Similarly, as instrument and operation faults cannot be completely avoided at plant site, the implementation of advanced tools for fault management (perhaps including the in-line control reconfiguration) must also be required during real ILBC implementations, as the process operation must be continued in spite of the faulty event, such as leaking, valve sticking, instrumentation failure, and feed contamination. As a consequence, protection actions and protection mechanisms must be employed to support the ILC decisions and the mechanisms used to maintain the product certification scheme.

THE ILBC UNIT OPERATION

The ILC procedure implicitly breaks the paradigm of the batch operation and, by doing so, suggests a newmodus operandi for the process operation. It is expected that stream properties in process lines present larger variability, when compared to bulk properties of stored products. Therefore, for the sake of certification, this new vigorous dynamic operation status certainly requires tighter control. However, the proper in-line measurement of the whole suit of fuel properties, necessary to certification, is one of the two major technical challenges. The other one is the analytical accuracy and reliability.

Efforts have been made to address the control problem, but the in-line instrumentation performance has seldom been compared to the off-line standard test ones for the sake of product certification (using repeatability and reprodutibility); correlation measures are provided instead [4, 13, 14]. In this sense, data validation is the key to assure the measurement precision, enhancing the signals of the sensors. Since instruments must operate in-line for long periods of time, the long-term behavior of process instrumentation must be properly evaluated and characterized. In addition, part of the required properties may be not available in-line due to technological or economic issues. The remaining of this section is dedicated to examine expected problems related to the ILBC low-level operational layers of the superstructure schematized in Figure 4.

Measurement

Before reviewing the main techniques and instrumentation used to monitor fuel properties, two major practical issues must be discussed: sample representativeness and sensibility to measured properties.

Sample representativeness is especially critical when analyzers require long times to perform the analyses. More specifically, most property analyzers do not operate on a real-time basis since the analyses are complex and time demanding. The product flow rate and the sample size must be defined in such a mode that the on-line analysis shows volumetric compliance without incuring into any constraint on the system production.

Addressing in-line measurement, ASTM D4177 standard focuses on practices for automatic sampling of petroleum and petroleum products whereas the ASTM D6624 standard is concerned with practices for determination of average property values of product streams (FPAPV). Lastly, the ASTM D7453 standard accounts for practices for sampling petroleum products by process stream analyzers.

Another important question refers to the analyzer sensitivity to a particular property. Such a question has been extensively studied in the literature, for example, using NIR technology [13], and different commercial solutions addressing this issue have been released in the market, for example, the FUELex analyzer of the Bruker Optics company. In fact, the search for surrogated variables and correlation development are encouraged. Agoston et al. [15, 16] introduced alternative sensors which enable in-line oil viscosity and corrosiveness monitoring, respectively. Such sensors surely agree with the ILC-based operational philosophy. The ASTM D6708 standard defines some statistical procedures used for assessment and improvement of the expected agreement between two test methods that are assumed to measure the same property of a material.

Concerning the technology used for in-line measurement, in-line specialized analyzers constitute the first reasonable option, at least for simple properties, such as density and conductivity. These analyzers are usually cheap, reliable, and fast, making them appropriate for composing an ILBC scheme. For properties hard to measure in line, inferences based on combinations of electromagnetic spectroscopy and inference models have become very popular both in academic and industrial applications [13]. The combination of NIR spectrometers and multivariate calibration techniques certainly constitutes the most popular alternative for inference development due to fast response times and high precision.

In general, NIR systems present better precision (repeatability) than ASTM analyzers and same accuracy (reproducibility) as the calibration method [17] and can produce one complete set of quality determinations in 30 seconds [5, 8]. Real-world NIR in-line applications have become very attractive, since 10 or more product qualities can efficiently be evaluated with NIR technology (diesel or gasoline) herein its ability to predict blend indices which are extensively used in off-line applications, as refinery production planning and scheduling. Practices for infrared multivariate and quantitative analyses are addressed by ASTM E1655. On the other hand, while equipments based on nuclear magnetic resonance (NMR) have also been investigated with comparable performance [18], less applications have been verified.

As pointed out by some authors [19], NIR-based analyzers require careful building of the inferential property model. However, suppliers of NIR-based analyzers offer poor modeling support and do not provide guarantees for measurement performance [20]. Acceptable performance assumes that the defined ASTM precision is reached at least 90% of the time during normal operation with no more than two model updates every year [20]. Other common complaints include filters clogging and product film deposition on optical parts. Proper instrument installation standards are described in ASTM D6122, which recommends sample conditioning systems design including automatic probe wash and regular model validation (daily or twice a day) [20].

Alternative manners to cope with the aforementioned problems are based on identification and analyses of the actual sources of property variability: the sample composition [21]. It is well known that the composition of a mixture can drive its chemical and physical behavior [14]. Recently, a promising branch of analytical chemistry based on advanced mass spectrometry, usually named FT-ICR-MS or simply FTMS [21], has been applied to determine the fuel properties based on a detailed composition of the sample. This has led to development of a new family of analytical techniques called petroleomics.

Although these techniques are still incipient and practically restricted to laboratories, typical analyses last from 5 to 15 minutes, making feasible in-line implementations. More mature techniques that can be used to provide the detailed composition of oil products include bidimensional (2d) chromatography, especially GC×GC and GC×MS [22]. However, the main disadvantage in this case is the trade-off between data resolution and the time required for analyses, around 80 minutes. Specifically, chromatographers can utilize very different columns, with different lengths and diameters, which influence resolution, analysis time, maintenance (clogging), and so forth.

Since the abovementioned equipments measure compositional variables, it is necessary to implement particular knowledge to convert such data into specified properties. In this subject, quantitative structure-property relationships have shown very promising results [23, 24], despite the sparse data of pure compounds. Table 1 shows a summary of the measurement technologies discussed here.

Table 1: Technologies to integrate an ILBC-based operation model

Technologies	References	Pros	Cons
Spectroscopy	Gilbert et al., 2004 [18] Morris et al., 2009 [13]	Fast analysis; high precision; low prices	Subject to local and periodic calibrations; insensitive sulfur compounds
(2d) chromatography	Yang et al., 2002 [25] Wang et al., 2005 [22]	Compositional data; established technology; varied configurations	Trade-off between analysis time and data resolution for real-time applications
Mass spectrometry	Marshall and Rodgers, 2004 [21]	Fast analysis; ultrahigh resolution; detailed characterization;	Qualitative data, due to ionization deficiencies; high prices
Quantitative structure-property relationships	Creton et al., 2010 [23] Saldana et al., 2011 [24]	Fundamental knowledge; universal applications	Few pure compound properties known; small data bases

A plausible model for producing a combined CoA may be based on Figure 7 schematics, where the NIR predictions are first generated during in-line blending and the QSPR predictions are then generated, using the compositional analyses carried out by chromatographers or spectrometers.

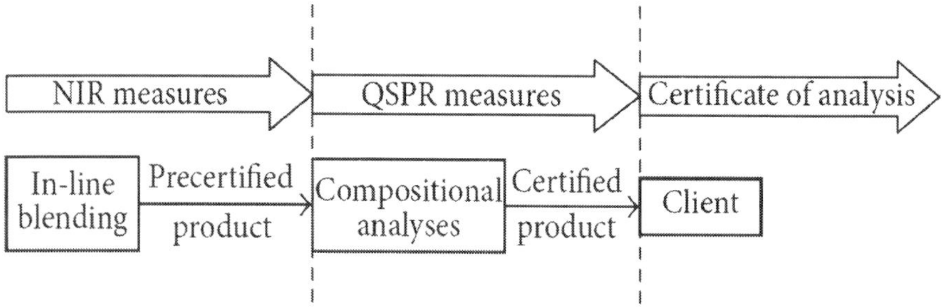

Figure 7: Basic ILBC analytical steps.

Validation

At the second level of the operation hierarchy, validation of obtained measurements must be performed in order to assure product certification. Such control layer enhances the measurement quality by identifying outliers, noise, or data trends and, if necessary, performing data reconciliation. Moreover, in case of identification of off-spec materials, protection mechanisms must detect and execute the necessary actions to ensure process safety and product quality.

Data validation is usually carried out by characterizing the measurement error, frequently measuring protofuels. Alternatively, results obtained by on-line analyzers can be compared to standard laboratory results according to statistical techniques aiming at detecting and quantifying discrepancies. The main standard practices available for validating on-line analyzers are ASTM D3764, for performance validation of process stream analyzers, ASTM D6299, for statistical quality assurance and control charting techniques of measurement systems, and ASTM D6122, for validation of multivariate on-line, at-line, and laboratory NIR-based analyzers.

Due to well-known limitations of conventional validation practices, which are usually performed with long time intervals, the measurement reliability is not ensured during the whole period of product transferring. Besides, precision and accuracy of nonconventional on-line analyzers are insufficient for product certification or have not been characterized. In this scenario, simple multivariate techniques, such as principal component analysis (PCA) and canonical variate analysis (CVA), can be of great interest to provide process data at short time intervals and reduce uncertainties, improving data repeatability and reproducibility [26].

Basically, multivariate techniques use correlations of properties measured by different on-line analyzers to increase data redundancy and to enhance capabilities for process fault detection. Abnormal behaviors can also be detected by comparing current data with successful historical runs. For example, NIR and FTMS analyzers can provide electromagnetic and mass spectra that can be used directly to provide operation states and monitor the process behavior [27]. In case of sensor malfunction or maintenance, multivariate techniques can be used to estimate the missing properties based on correlations with measurements provided by other sensors and/or historical models, allowing for acceptable ILC robustness [26]. In case of process fault, automatic protection actions must be triggered, such as segregation of the off-spec components or product, in order to guarantee the ILBC premises.

Figure 8 illustrates the worst validation scenario of the proposed ILBC scheme, in which all components behave abnormally, yielding an off-spec product that must be isolated from the jetty line. After problem detection with the help of electromagnetic spectra and petroleomic tools, the product is then sent back to an off-spec product storage tank through a recycle line, for posterior reblending or reprocessing.

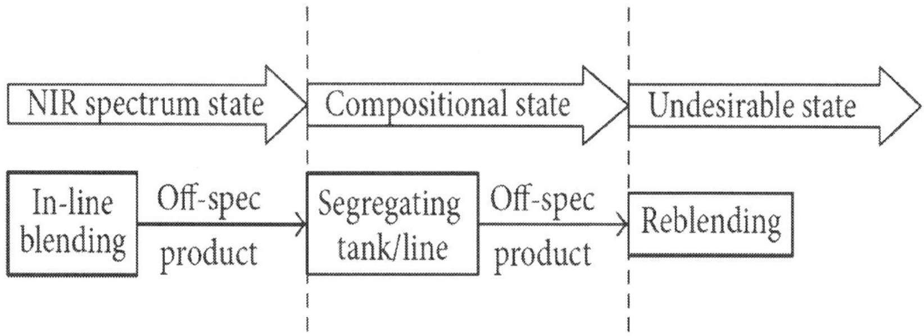

Figure 8: Segregating route of ILC procedure.

The Human Organization

Changing attitudes, revising previous assumptions, and rethinking best practices for refinery operation are particular challenges of all in performing optimization.

Actually, rather than an interdisciplinary activity in nature, ILBC operation must be understood as a very specialized optimization activity inside refinery. A major challenge to successfully implement an ILBC technology may indeed be educating the refinery's personnel to work according to new paradigms and work process.

Since at the refinery technical-commercial boundary, the ILBC operation is not only strongly dependent on the refinery production scheduling but also provides a vital feedback for its daily updating. Such collaborative scheduling formulation may be especially important for refineries embedded in a very competitive environment, in which scheduling is usually market oriented. In this sense, rather than any computational systems, the quality of the staff may be posed as the prime success factor, along with their ability to work as a single team, despite focusing on distinct problems and tasks [12].

Best work processes in turn enable the personnel to consistently achieve production plan objectives. Therefore, work process should be independent of personnel changes. This is accomplished via structuring a refinery team responsible for the ILBC unit performance. Some critical characteristics of a successful team are identified. As pointed out by Barsamian [5], thinking and working oriented to "do things right the first time" should drive local raining people's attitudes and mindsets aiming at supporting the ILBC unit. Rapid communications, faster response times (analyzer repairs with high priority), key people on call, and management motivation of ILBC team members are mandatory practices that motivate a matrix type organization in which the individuals remain physically in their respective areas, but their daily work responsibility is to the team (Figure 9).

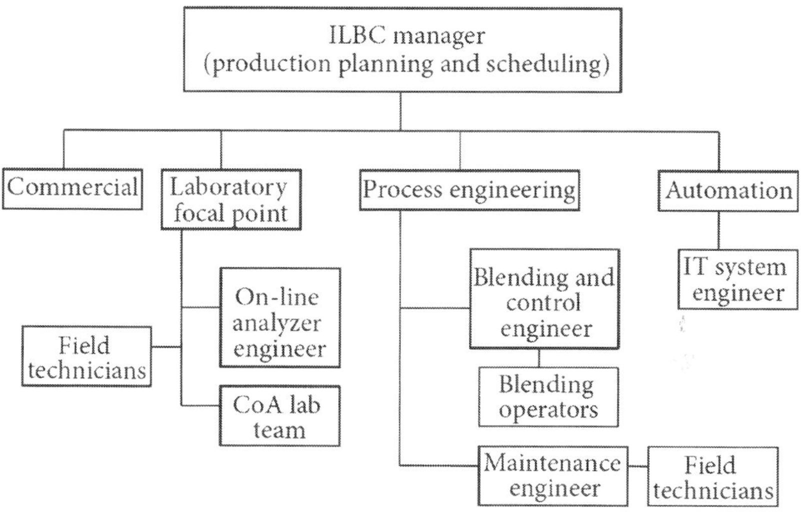

Figure 9: A ILBC team model [5].

The ILBC team is responsible for the daily operation of the ILBC unit. It comprises tight monitoring of performance indicators including critical analysis of the nature and frequency of operational issues and incidents, definition of priorities related to system maintenance and upgrading, resource forecasting to ensure high factors of system availability and reliability, and managerial follow-up.

The most prominent change in the refinery's personnel profile is certainly experimented by the laboratory staff. Not only a new work philosophy must be learned, but also a set of more added-value activities should be performed. Since now focusing on the on-line analyzers performance, the laboratory's staff must be qualified to locally provide process streams and fuel additives spectral modeling, protofuel management, research on physicochemical inferences development, and equipment monitoring and maintenance. As product certification process is expected to be less demanding for laboratory ASTM analysis of manual operational routines, as periodic line and tank sampling, mechanical work should be replaced by challenging activities performed by high-profile professionals. On the other hand, it is uncertain to say that ILBC automation will cause the reduction of laboratory activities. Anyway, it is worthwhile to note that those refiners that did not reduce their technical staff when automation was introduced have experienced gains in operational safety [28].

Since such technologies are a major concern as they impact the entire plant schedule and real-time optimization process applications, a highly qualified and integrated blending team becomes instrumental to the modern refining industry.

Superior Control Layers

The remaining control layers of the proposed ILBC operation hierarchy are concerned with the blending process operations. However, it is important to note that (a) product delivery underscored by ILBC technology is faster than the conventional one (Figure 2) and (b) on-line properties measurements present more vigorous dynamics and this way requires tighter control. Once the property mixture rules and indexes are known and the inferential models are implemented, NIR-based analyzers can quickly measure the feed and product properties [19]. Hence, it allows for a more sensitive control using both feedback and feed forward control structures. An important feature of on-line measurements of oil properties is the higher frequency of analyses, which enables improved process operation and more frequent optimization of process setpoints. For instance, both electromagnetic and mass spectra can offer essential information for definition of the best recipe for a particular production run [27].

ILBC PREREQUISITES

Custody transfer involves industry and legal metrology standards, contractual agreements between the parties, and governmental regulation and taxation. Therefore, any concrete investment decision on ILBC implementation should be preceded by an interdisciplinary analysis of the refinery and the environment in which it is inserted. Major issues are addressed in this section, as follows.

Technical Issues

At the technical viewpoint, an ILBC unit may be understood as a forward step of a reliable ILB unit. The success of an ILBC implementation is conditioned to the maturity stage of the refinery ILB automation technology as well as its previous experience in dealing with ILB operations. Table 2 summarizes major qualitative characteristics expected to further raise a successful ILBC application.

Table 2: Some critical technical requirements for an ILBC-based operation model

Discipline	Item	Prerequisite
Analytical	ASTM analyzers	Reliability
	On-line analyzers	RobustnessSample condition system Redundancy with on-line ASTM analyzers
	Sampling	Sampling conditioning systems (on-line measurements) Automated composite sampling
	Laboratory	Laboratory NIR spectrometer Reference materials/ protofuels Spectral database (protofuels, streams, and additives)

Information technology	Real-time database	Process streams data storage (extensive/intensive properties) Discrete and continuous plant data storage (e.g., on/off flows, tank heels)
	Off-line database	Process configuration (e.g., equipment operational limits)
Process engineering	System design	Allowing for redundant measurements
	Process modeling and control	Reliable inferences to noncritical product qualities MPC regarding feed forward on components blend indices Blending RTO technology Ratio control including fall-back strategies
	Oil movement system	Integration to refinery production scheduling Automated plant operation management (i) Tanks: swings, heel qualities, and stratification detection (ii) Line ups: flushing, valve position, and pipe quality tracking
	Data validation	Real-time data reconciliation on flows On-line analyzer data validation on ASTM analyzers

Legal Issues

The first question that must be considered prior to legal implementation of an ILBC procedure is the following: is it legally permitted?

Under the preceding US EPA proposition for shipping before testing based only on on-line analyzer control of properties of interest to EPA (USA Code of Federal Regulation, 40CFR80, part 65), governmental regulatory legislation worldwide should be expected to be revised in the light of the fast development of ILBC technology as well as due to environmental pressures, since it may optimize volatile organic compound emissions at the end of refinery. As a result, the formulation of new modus operandi and mechanisms to ensure consistent ILBC implementation and customer rights may be propelled by prospective studies and pilot projects by the oil industry.

Chapter 6

Rationalizing the Use of Water in Industry—Part 2: Instruments Developed by the Clean Technology Network in the State of Bahia

Asher Kiperstok, Karla Esquerre, Ricardo Kalid, Emerson Sales, Geiza Oliveira

Federal University of Bahia (UFBA), Clean Technology Network of Bahia (TECLIM), Industrial Engineering Program (PEI), Rua Aristides Novis, Salvador, Brazil.

ABSTRACT

The instruments developed by the Clean Technology Network of Bahia (TECLIM) at the Federal University of Bahia (UFBA) (cited in Part 1 of this paper) are presented. Factors regarding water management in industry were examined, on the basis of experience acquired over the period of a decade in cooperative research projects with large industrial process plants located mostly in the Camaçari Petrochemical Complex, Bahia State, Brazil. The main results consist of training about 1700 industry professionals in CP, the identification of about 500 ideas for the rationalization of water use, the presentation and publication of 90 articles in journals, conferences and other academic events, identification of ideas with potential water savings estimated at around 1400 $t{\cdot}h^{-1}$ and the reduction of at least 500 $t{\cdot}h^{-1}$ in effluents. Other sectors that make use of water, for example public buildings, commercial buildings, homes, shopping centers and airports can adapt and use the TECLIM method as will be exemplified.

INTRODUCTION

As cited by Kiperstok et al. (Part 1) [1], during the course of cooperative projects with industry, the Clean Technology Network (TECLIM) has developed a method for the optimization of water use in industries. This method is based on CP principles and is composed of some of its own instruments and others adapted from literature [2-5]:

1 The merging of academic and operational/Industrial knowledge;
2 The inclusion of CP concepts by means of continuous large-scale training;
3 Construction of reconciled aqueous stream balances considering uncertainties in measurements;
4 Compilation of a digital database of ideas that takes into account cultural, environmental and economic factors in evaluating the potential use and drawbacks of opportunities;
5 The introduction of a geographical information system (GIS) to identify the sources of water (by producers or consumers) within the context of companies and in the region;

6Optimization of Mass Exchange Networks;

7 Analysis of the region's hydro and hydro-geological characteristics;

8 Drawing up conceptual projects for reducing the use of water and generation of effluents;

9 Auditing of the sources that give rise to the effluxents.

These instruments are described in the following sections considering their respective goals and results.

INSTRUMENTS

The Merging of Academic and Operational/ Industrial Knowledge

University-industry partnerships in the field of science and technology aim to transfer the knowledge generated within the universities to industries with a beneficial impact for society in general and local business/industry in particular. In these cooperative projects, the university increases its contribution to research through public and private financing and experiences current practice in industry [6, 7].

There are clear differences between the culture of industrial and academic environments. Each has their own distinct objectives and working conditions, both in general terms and in the ways they seek to bring about environmental sustainability. Whereas the industrial sector faces situations that have to be deal with short or medium terms, with regard to creating, following and consolidating current practice and procedures, the academic world designs projects to be undertaken over a longer period of time. While the industrial sector requires immediate action, the academic world devotes time to formulating new theories. Evidently, bringing these two kinds of knowledge together can be of immense value to both areas, as there is already a great deal of mutual respect and willingness to achieve common goals [3].

To lower the barriers between academia and industry, a balance between academic goals and industrial priorities, involvement and commitment of the partners, mutual trust, and effective communication are the key factors. The quality of project management must respond to external changes or strategic addition of large doses of patience and hard work [8,9].

In the Industry-TECLIM partnerships the analysis of the water management in each company consists of an initial constructive stage for each research project. The objectives and expected results can thus be defined by taking this examination into account, together with the instruments that have been developed.

The team responsible for undertaking each project (called Eco-team) comprises graduate, post-graduate, master's and doctorate researchers in chemistry, and in chemical, electrical, mechanical and environmental engineering, as well as engineering trainees. This Eco-team is supervised and supported by a team of coordinators (university lecturers) and facilitators from the industry.

In general terms, the activities can be divided into two distinct stages: In the first stage, the researchers concentrate their activities at the university and seek to gather information about the processes and historical background of the use of water and the generation of effluents. In the second stage, when the researchers have gained a degree of autonomy, they begin to work inside the battery limits of the industries. Computer-based support tools are used in the planning phase. This process is periodically kept under review and updated to take into account new information and the guidelines developed through the experience of approaching academic and operational knowledge.

Some managers and partners from both sides take part in an Advisory Council which is set up to monitor the preliminary policies and offer guidance to both the preparatory phase and the progress of the studies. This council plays a key role in the allocation of the company's resources, mainly the work schedule of participants on the project, as the allocation of the funds are generally dealt with by the team leaders.

Another kind of interaction between the university and companies occurs through weekly meetings that are held with researchers and facilitators from the industry. These meetings are designed to oversee the progress of the studies, make recommendations and set out new actions. There are also less frequent meetings with the Advisory Council which, in turn, takes a close look at the suggestions made by the researchers involved in the project in the operational areas under study. In a project lasting two years, for example, the Advisory Council usually holds six meetings.

The monitoring of the environmental performance of the company with regard to its water consumption and generation of effluents is carried out by means of indicators dealing with the water consumption, the amount of effluent generated and the rate of production in the same period. The causes of the variations of these indicators are analyzed as a part of an optimization strategy to determine levels of water use and the reduction of the generation of effluents.

It should be stressed that 50% to 60% of the time spent on the project is invested in the collection, compilation and validation of the information collected.

Insertion of the CP Concepts through Continuous Large-scale Training (I2)

To implement CP principles in industry all the employees should be involved, if possible [10]. Successfully implementation of this practice requires also appropriate training of all the Eco-team from the industry and the university [11].

The guidelines laid down by the Cleaner Production (CP) are used to oversee the work carried out in each company and to identify the causes of unnecessary water usage and the generation of effluents inside the battery limits (and not just on the margins). The purpose of this activity is to heighten the awareness of the operators and engineers about how water is managed inside the Industry. It requires stopping the delegation of issues concerning water usage and the generation of effluents to third parties.

The personnel are placed at the forefront of the challenge of creating environmental sustainability by employing eco-efficient factors (Factor 10 and Factor X, according to Graedel and Allenby [12]). This approach highlights the inefficiency associated with the end of pipe solutions and stresses the need to find a solution that addresses the core of the problem—the source of the water waste. Only after reducing the problems at the source can new ways be sought by the Eco-team for the reuse and recycling of effluents. In all circumstances, effluent must be regarded a fluid that forms part of the production processes despite its inferior quality. Alternative forms of treatment and disposal may also be considered, but only as a last resort.

The complexity of the processes and the lack of and often poorly organized information available are among the difficulties to be overcome, thus requiring both structured and systemic method of data collection and management. The inclusion of the CP concepts presupposes a change in the culture of the company. How can the cultural barriers be broken to allow the CP concepts to be effectively incorporated? How can people be dislodged from their respective comfort zones and moved to a region where they must learn and create? Changes of such magnitude may arise from the knowledge of the flows handled by the operators, structured by a water balance, and with large-scale training of professionals through a culture of continuous learning.

In the developed method, the concepts of CP are usually taught in courses lasting for 12 hours and running in 2 stages. The course is offered to the whole group of operators and engineers, although it can and should be extended to all the company (including the third-party companies).

In the first stage of the course, an attempt is made to increase the awareness of the participants in matters regarding the consumption of natural resources and to shift their attention from the so-called end of pipe technology to the source of inefficient water use in the upstream and downstream processes themselves and their relations with other processes in the productive sector. There is a review of a conceptual and simplified version of the method for

CP offered by UNEP/UNIDO/CNTL [UN Environment Program/ UN Industrial Development Organization/National Center for Clean Technologies] (FIESP—Federation of Industries in the State of São Paulo [13]). After dividing the participants into assorted teams (reflecting different sectors of the process) of four or five people, each team is encouraged to develop a project by using CP concepts to find new ways to improve the environmental performance of the company, including in the operational and administrative areas.

The teams have a one-month deadline to carry out this project, and then they show the results in the second stage of the course. The researchers on the project also attend the course and support the teams by structuring the new opportunities that have been found. After the teams have explained their ideas and suggestions, environmental, operational and economic factors are raised such as the area and equipment involved, the estimated savings of water and the funding required for their implementation.

The ideas are registered and assorted in a digital database of ideas (see instrument I4) which is periodically analyzed by the Eco-team to determine the technical and financial feasibility of these ideas and whether (after examining them in greater depth) they might lead to conceptual projects which can be carried out by researchers together with a specialist team of technicians from the company. This training fosters each participant to seek eco-efficiency in the workplace as well as in their own private lives.

Apart from the CP course, the members of the Ecoteam undergo an intensive training sessions which are given at the beginning of the project. These courses and lectures are open to the whole university and to every company employee. Table 1 shows a list of the main courses generally undertaken.

Table 1: List of training offered in the beginning of the projects

Key topics covered in training	Hours	Count	Lecture
Flow measurement. metrology and uncertainty	40	X	

Simulation using EMSO	18	X	
Environmental statistics	18	X	
Project of network to industrial water reuse	17	X	
Software MS project	16	X	
Cooling towers	8		X
Water balance and uncertainty incorporation	7.5		X
Mass Exchange Network and RecoVR software (Application for water resource management and network design reuse)	7	X	
Cleaner production	6		X
Overview of industry production process	4		X
Introduction to MS project	3.5		X
Water quality and equipment for water and wastewater treatment	3.5		X
Data reconciliation without redundancy of data measurement	3		X
Reading and intapretation of calibration reports	3		X
Water resources of the metropolitan region of Salvador (RMS)	2		X
Carat expression measurements and estimates	2		X

Water Balance and Data Reconciliation (I3)

The lack of measurement of water flows in processes is a constant source of problems in industrial plants. This arises either because there is a lack of measuring devices or from the fact that they are not handled correctly. Both cases reflect the low priority attributed to water management in industry. This scenario has been addressed by constructing water balances using the few measured aqueous flows, and the other values being estimated.

This instrument represents the rate of the aqueous flow at the entrance and exit points with an approach that involves various control volumes (for example regional, industrial plant, areas and units of production).

The study of the mass flow of aqueous streams either in the form of water supply (untreated, clarified, demineralized and drinkable), steam (under various thermodynamic conditions), condensate (clean or contaminated) or effluents (of the most varied kinds) allows the team to trace a water profile in the factory and thus enable improvements in the management of this resource. As a result, all kinds of data about the flow can be used whether they originate from measuring devices which are rarely calibrated, or estimates of the flow obtained by means of mass balance, effective spot measurements, project data, theoretical estimates, simulation or even information collected in the unit through interviews with the operators. This search for information together with the involvement of the company's personnel to find ways to reduce water use constitutes one of the most important stages in the whole process.

Each source of information is assigned a score (QI)— the degree of reliability or the Quality of Information. A macro-perspective can be obtained from the global Water Balance. This covers the most comprehensive control volumes at the entrance and exit points of the industrial plant, see Figure 1. More specific perspectives can be obtained by identifying the currents at the entrance and exit points of each unit or piece of equipment.

All the data in the Water Balances (viewed in general terms and per unit of operation) are compiled in electronic spreadsheets and represented in the form of a block diagram. Each block represents one of the production units or even each kind of equipment (identified through its Business Identity Code—ID or TAG number).

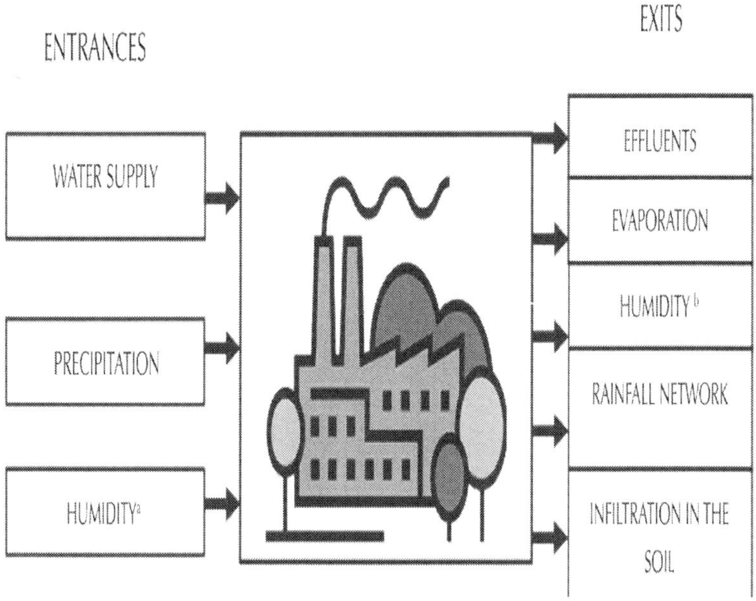

Figure 1: Global Balance—the main entrance and exit currents. Source: Kiperstok et al. [2].

Each connecting line in these blocks indicates the aqueous currents that flow in or out and the arrows indicate the direction of the flow. The lines are represented in different ways depending on the type of current-dashed, colored or with an acronym in the caption.

To compare the currents, the data are represented on a continuous base-line in t/h. A quality of information (QI) score is linked to each part of the information about the flow. Information uncertainty, which is in inverse proportion to the QI, can be attributed to the method used to obtain this information. Initially, the QI values between 0.4 and 10 are given for each aqueous current under consideration [5] where the score of 0.4 refers to a rough estimate and a score of 10 is given in cases where there is a calibrated measuring device (Figure 2).

One possible way of assessing the value of the global QI of the various control volumes is to make a calculation of a representative

QI of the amount under consideration by weighing the values of the flow and the QI of the known currents (Equation (1)).

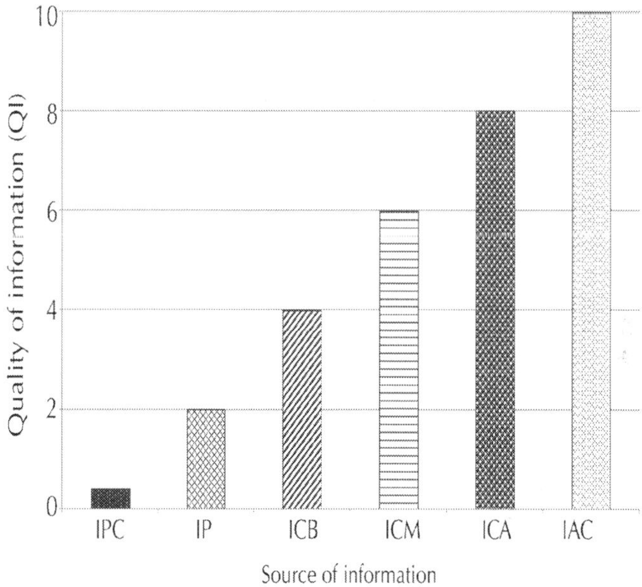

Figure 2: Attributed values for the Quality of Information (QI) of the rates of flow that are taken into account in the Water Balance. Source of Information: UI: unreliable information—based on rough estimates; PI: information of slight reliability—simulation and/or based on old projects and rough measurements; ILR: information of low reliability based on operational and descriptive records of outdated processes; estimates made by engineers or operators who have worked in the unit or based on a device used for detecting lack of calibration; IAR: information of average reliability—information supplied by engineers and operators with little experience and metering in normal operating conditions; HRI: highly reliable information—information which is validated by the operation and/or engineering and current simulation or project, and with a history of metering in normal operating conditions; ERI: extremely reliable information—measuring with a calibration device, simulation of the processes validated by the engineer and information supplied by engineers and operators who have a good deal of experience and are still active in the area.

$$QI_{ponderado} = \sum_{i=1}^{N} QI_i \cdot V_i \Big/ \sum_{i=1}^{N} {}_i \tag{1}$$

where QI is the quality of information attributed to the value of the flow i; V_i is the value of flow of the water current i and N is the number of streams considered.

One way to improve uncertainty distributions and ensure that the information complies with the determined restrictions imposed on the process (Deviation = 0) consists of applying the data reconciliation of the mapped flow rates [14]. On the basis of the flow data associated with its quality (QI), data reconciliation seeks new flow values that can satisfy the mass balance equations by means of mathematical optimization techniques. This operation reduces the difference between the original values and the reconciled flow values.

The difficulty in using a typical formulation for the problem of data reconciliation lies in the need to know the uncertainty of the estimated data [15]. To overcome this difficulty we developed a relationship between QI and the uncertainty of the data [5] as shown in Equation (2).

$$\sigma = k\left(VM_i / QI_i\right) \tag{2}$$

The k parameter is constant in a water balance, so the objective function of the data reconciliation can be rewritten in terms of QI and without this parameter, see Equation (3).

The TECLIM/UFBA method suggests a way of circumventing this problem by using the formulation outlined in Equation 3 that regards the QI as having a behavior that is the inverse of uncertainty. The problem of optimization can be solved, for example, by using the optimization toolbox from the MATLAB software and introducing the flow data and the QI that is available in the electronic spreadsheets. This stage of the data reconciliation is essential to increase the reliability of the mass balance.

$$\min \sum_{i=1}^{N} \left(V_{M_i} - V_{R_i}\right)^2 \cdot \left(QI_i^2 / VM_i^2\right) \tag{3}$$

Where: V_R and VM are the flow rates that are reconciled and mapped (or presumed to be mapped) with reference to current i; N is the number of currents involved.

After reconciliation the uncertainty and QI can be evaluated by propagation on uncertainty [16,17].

Implementation of a Bank of Ideas (I4)

Environmental optimization of industry involves extensive work in identifying opportunities to improve productive processes and thus the management of these opportunities becomes of decisive importance when making use of the creative potential of the participants and collaborators. A Bank of Ideas was developed to allow a continuous flow of records to be kept for subsequent analysis of prospective ideas/suggestions facilitating the participants and collaborators from the companies to become involved in the process of rationalizing water use.

The instrument is set within the vision of the method Front-End Loading (FEL) for project management. FEL was developed by IPA (Independent Project Analysis) and is mainly applied in investment projects in order to reduce the risks involved in decision-making as well as costs related to reworking and redesigning [18,19]. This method is applied as a lever in the early stages of a project (initiating and planning) that directly impact the success of the project. The Bank of Ideas in TECLIM's Method was developed with the same goals as FEL; to increase the likelihood that only the ideas that are well developed become conceptual projects (Instrument I8).

The structure of the Bank of Ideas involves a database in which the information of each idea and user can be stored. The operation of this database can be divided into two periods. The first involves the registering of suggestions by the users and the second the administrative work carried out by the Eco-team. The users are asked to register their ideas and carry out periodic check-ups, while relying on the Eco-team to help them to proceed with their studies. In contrast, the Eco-team has to make a disclosure of the contents of the database, correct any mistakes in the ideas already registered, encourage and publish an assessment of the ideas, and seek sources that can lead to improvements to make each idea more viable.

In the evaluation stage, the ideas that have the greatest potential to be turned into opportunities are identified. This is done by sorting the ideas into the different Industrial areas to which they are linked. At the same time, the Eco-team selects employees from the company to assess the ideas that correspond to their respective areas. When the idea concerns more than one area of the industrial plant or can be applied to the whole industry, the group of assessors is composed of researchers who are involved in the project and company employees that have comprehensive knowledge of the productive process. An evaluation form was structured to give support to the stage of assessing the ideas suggested. In undertaking this, three aspects of each idea are assessed:

- Environmental factors: regarding the reduction of water consumption or wastewater generation, for example;
- Cultural (or psycho-social) factors such as changes in attitude and behavior both within and outside the industrial plant;
- Economic/financial factors: the estimated financial return that can be made on an investment by putting a new opportunity into effect.

Table 2 shows an example of the percentage bands and approximate values (in R$) which are used to carry out the assessment of the respective environmental, cultural and financial factors. On a scale of 1 to 5, the ideas that are technically the easiest to implement and that can have the most beneficial effect on environmental Industrial relations are given the highest scores.

The weighted averages of the scores assigned to each idea by the evaluators are calculated and shown in the graph displaying the ranking of opportunities, see Figure 3.

The ideas located in the upper left-hand square are, theoretically, those whose implementation would bring about the most positive effects with the least financial investment being made, suggesting that they should have priority for study. As for the ideas placed in the upper right-hand square, these require more funding, although they show the same positive effects. In contrast, the lower left-hand square brings together the ideas that require less financial investment and have a less attractive effect while in the lower right-hand square, the ideas are not appealing. This stage completes the final objective of the management of the Ideas Bank: to identify the ideas that have the best prospects of being turned into opportunities.

The Geographical Information System—GIS (I5)

The spatial representation of the area under study is of crucial importance in giving assistance both to decisionmaking on how to use the aqueous flows at each point of the process as well as to design strategies to improve the use of the hydric resources available in the surrounding area of each company and region.

The Geographical Information System (GIS) is introduced with the aim of providing an integrated view of the area where each project takes place and, more specifically, of forging links between the sources and consumers of water, which are represented by geographical coordinates. The purpose of this instrument is to serve as a basis for making an assessment of the costs incurred in linking up the effluent flows with their destinations.

Important information for water reuse modeling includes: equipment and process unity location (to determine the distance between points), relative elevation (to determine the need for pumping) and water properties such as quality at the input and output of a particular process or equipment. Based on this information GIS creates possible paths of exchange, which will be evaluated by the Eco-team for their ability to minimize water and wastewater costs and reduce environmental impacts [20].

Table 2: Scores to evaluate the positive effects of an idea

Score	Concept	Initial investment an-doperational invest-ment (RS. Year[1])	Positive Impact		
			Environmental	Cultural	Financial (RS. year')[a]
			Weight= 5	Weight=2	Weight = 3
1	Very low	>5000	<0.1	Post	>5000
2	Low	(5000: 50.000)	(0.1:0.4)	Area	(5000: 50.000)
3	Average	(50.000:500.000)	(0.5: 2.0)	Plant	(50.000: 500.000)
4	High	(500.000: 1.000.000)	(2.1: 4.0)	Plant + house	(500.000; 1.000.000)
5	Very high	>1.000.000	>4.0	Plant + house + community	>1.000.000

[a]The range of values should be compatible with the total assets of the industrial plants.

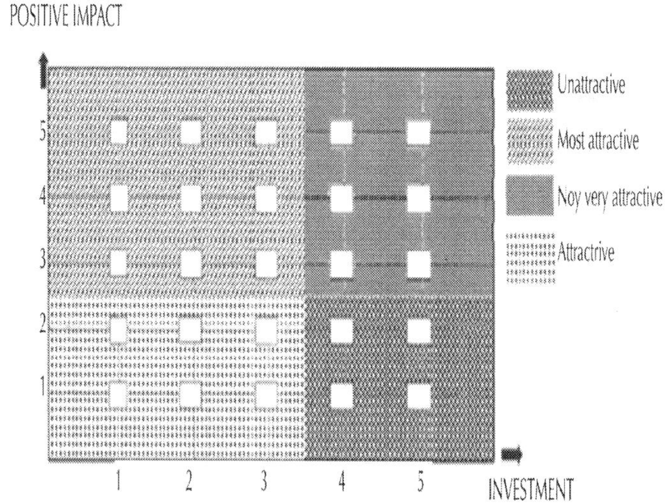

Figure 3: Graph showing the ranking of opportunities. Source: www. TECLIM.ufba.br/ideia.

The system is also capable of determining which localities are the most suitable for the installation of equipment that can restore the quality of the aqueous currents being reused. This instrument allows an exploratory analysis of information to be undertaken through an aggregation of spatial representation and its respective attributes of alphanumeric data linked to a database. Furthermore, it may allow new scenarios to emerge and thus facilitate decision-making.

The construction of an Information System Platform for the industrial management of water and effluents can be divided into three stages:

- The reproduction of the real world (the companies) in an analytical virtual world: characterization of the companies based on the productive sources of water and effluents, the points of water consumption, and the particular hydro and hydro-geological features of the locality and the region.
- The setting up and utilization of the database with information gathered from the company or even external sources of data, including information concerning the water availability of the region.
- Identification and analyses of the opportunities regarding rational use of water and the reduction of the generation of effluents, as well as integration of the data from the conceptual projects proposed from the application of the instruments I4 and I8.

After this stage, GIS becomes a constant feature and every new piece of data should be incorporated. This instrument may be used as an aid for the water management and/or monitoring system

Optimization of the Mass Exchange Network (I6)

Water consumption reduction practices when dealing with a great amount of streams generally imply the synthesis of Mass Exchange Networks (MEN). These either use generalized optimization methods [21-24] or the water pinch analytical method, as outlined by Alva-Argáez et al. [25], Smith and Petela [26],Wang and Smith [27] and Ruiz [28].

According to Smith and Petela [26], the traditional procedures for reducing water use, such as changes in washing systems, have achieved a reduction of 30% - 60% of untreated water consumption in industrial applications when complemented with the water pinch method. An alternative method makes use of the Water Sources Diagram—WSD [29,30] to reduce the consumption of water and generation of effluents. In both cases, most of the time, some additional equipment is required to treat some streams to allow their reinsertion on the process.

The objectives of the optimization of MEN consist of developing and applying technologies, and methods that can make industrial processes more eco-efficient. This is achieved by using synthesis and optimization tools, economic engineering, and thermodynamic principles of mass exchange to design MENs with the minimum use of natural resources.

After constructing the water balance, the information is compiled in a table called a Challenge Table, which contains a detailed list of the information of every stream taking part in the hydric balance.

The following information can be obtained from the Challenge Table, after using techniques for integrating the processes: the required capacities for the treatment operations of the effluents, an estimate of the financial investment needed, and an estimate of the reduction in water consumption.

In the early stages, research groups were invited to work directly with optimization by using a Water Sources Diagram (WSD) [31]. One of the reasons for this was to make it easier for the engineers in the plants to understand and to adopt the method.

Some work has been carried out with the aid of the WSD methodology, including some Master's degree dissertations under the combined supervision of the lecturers of the Clean Technology Network (UFBA) and academics from outside UFRJ (Federal University of Rio de Janeiro) [30]. Despite the difficulty of working with real cases, with multi-component streams, poor measuring devices and historical data that have not been consolidated, the application of the method can identify a wide range of opportunities to reuse intermediate streams or other external sources of water. This can increase the efficiency in water use, if the final recommendations are put into effect.

Complete development of the MEN synthesis has not yet been reached for any of the industrial plants where they have been put into effect. The three main reasons for this are as follows: a lack of proper information regarding the aqueous flows at the level of the process, difficultties in attributing values to the maximum concentrations of a particular pollutant that has to be taken into account in the water that enters the processes, and finally, the fact that the methods that are applicable to the multicomponent processes are still at an early stage of development. Despite this, some recommendations were implemented to a lesser extent, with success in some of the TECLIM partners' sites.

Analysis of the Insertion of the Company in the Regional Hydrological Cycle (I7)

The main purpose of the studies undertaken regarding the influence of the company activities on the regional water cycle is to assess the effectiveness of measures taken for the protection, preservation and remediation of surface and subterranean water resources in the region. These activities take into account the assessment of the hydrological and hydrogeological features of the region and make it possible to set out the extent of the beneficial effects that any reduction in water consumption and generation of effluents can have beyond the boundaries of the company.

In addition, conducting a survey of the potential supplies of water in the region can enable the Eco-team to draw up strategies based on the Life Cycle Analysis (LCA) to improve water usage, taking into account its quality and the logistical factors involved in its use, as well as the need to find alternative sources of water supply and effluent disposal in the area, if needed.

Conceptual Project Design (I8)

The conceptual projects reflect the results of incorporating P + L concepts in the company and are based on the ideas registered in the Database of Ideas (Instrument I4). They undergo a preliminary analysis and take into account the spatial representation of the area under study (Instrument I5). At this stage, a multidisciplinary team is desirable, together with a training scheme, to deal with the different technical requirements of each company in preparing a conceptual project design.

Among other activities, these requirements involve working out strategies for introducing the idea, preliminary economical/financial feasibility studies, Hazards and Operability Analysis (HAZOP) and Failure Mode and Effects Analysis (FEMEA). Obviously, a review of the data regarding the consumption of utilities and raw materials, the effluents generated and the disposal systems is essential.

The schemes for the reduction and reuse of water and effluents within the units of the production process are drawn up based on project data, operating manuals, stream data from the Water Balance and the quality of the water and effluents found in the Challenge Table, as well as the information derived from various discussions with engineers and operators. As the aim of the TECLIM method does not include the development of detailed or engineering projects, it is usual to remain in contact with collaborators from the company for assistance.

Auditing Sources of Effluent (I9)

The aim of auditing effluent streams is to define actions and procedures to reduce the volume of both organic and inorganic effluents within each process unit. Another objective is to improve the quality of effluent by minimizing the undesirable loss of products through these streams.

The method uses the techniques mentioned in Avila [32,33]. One expected result from the application of this method consists of increasing the load and continuity of the production process due to its higher stability. It is also expected to improve the Eco-efficiency with a reduction of losses thereby improving the company's image internally and externally with increased customer satisfaction

RESULTS

General results in terms of reduction in water use and effluent generation were shown in the first part of this paper [1]. Further specific benefits obtained through the application of the instruments described here were presented. At least 90 technical articles directly related to the projects have already been published in journals, conferences and other academic events.

The estimated environmental benefits in terms of water and wastewater reductions based on the identified opportunities are summarized in Table 3.

The results shown in Table 3 consist of training about 1700 industry professionals in CP, the identification of about 500 ideas for the rationalization of water use and identification of ideas with potential water savings estimated at around 1400 $t{\cdot}h^{-1}$ and reduction of at least 500 $t{\cdot}h^{-1}$ of effluents.

Table 3: Environmental benefits related to each research project

Project	No. of profes-sionals trained in CP	No. of ideas gener-ated	Total water consump-tion (t.h^{-2})	Water use reduc-tion (•h^{-1})* and % about withdrawal flow	Wastewater reduction (t.h$^{-\prime}$)′ and % about original wastewater
A	50S	103	600	15$ (-26%)	33 (-18%)
B		53	4300	390 (-9%)	350 (-40%)
C	33S	113	407	160 (-3M)	NV
D	194	50	90	32 (-35%)	4 (-90%)
	400	106	1250	320 (-21%)	NV
F	150	67	700	300 (-40%)	120 (-45%)
Total	-1700	-500	-7400	-1400	-500

[A]values may change in ±20%; NV = not valued.

The attribution of the Quality of Information (QI) is a means itself of improving the efficiency of the process, and helps to establish priorities during the optimization step. The water balance has helped in the identification of the water flows (and their respective QI) which has to be considered for improvements an each process unit. In mapping out the aqueous flows, for example, company B found that 20% of them had not been taken into account by its information system [34]. With regard to the data reconciliation of the water balance, the big difference between the method developed by TECLIM and traditional methods is that the divergent information is not just eliminated from the objective function but is considered for a re-evaluation of its QI. This approach allows the data reconciliation to occur without any redundancy of measurement, which is normally required in the composition of the water balances.

The reconciliation of the water balance has already revealed serious errors of measurement that were not noted by the measurement systems of the companies being studied. For example, an error of 10% or more in the measurement of a flow of effluent sent to a treatment plant in a calibrated instrument was found. This value was of significance to the company owing to the volume of effluents generated (about 500 $t \cdot h^{-1}$) and the considerable cost of the treatment. Another failing identified by the reconciliation was in a company that had operated for many years on the basis of two aqueous currents that had been mixed up. The reconciliation revealed this error and discussions between the UFBA team and the company staff led to the discovery of the existence of a mixed measurement in the DSDC (Digital System of Distributed Control) following an inspection of the system.

Apart from the data collection and the attribution of the quality of information, attempts have also been made to tackle the problem of finding the best effluent allocation such as the water used in the process, by employing the geographical information system.

GIS enables engineers to make a quick assessment of the necessary logistical issues (distance, quotas, removal of obstacles, free passage) for mass exchange from one point of the plant to another. In this way, ideas that are less viable can be swiftly rejected. GIS was also used to define the areas that could benefit from the use of collected rainfall and it also identifies the companies that are impacting the water cycle of the region where they are based more. This instrument was very useful in all the projects carried out.

The analysis of the way the company is incorporated in the hydrological cycle led to a reduction in the water withdrawal from the water sources that supply the Industrial Complex where most of the companies being studied are situated. The search for alternative water supplies in this complex led to the identification of three contention basins of rainfall and inorganic effluents [35,36] and a hydraulic barrier system with 9 extraction points of contaminated ground water [35] that are potential alternative water suppliers.

These alternative water sources offer ways of managing the water flow to the companies by broadening the forms of supply and releasing it from its dependency on the actual water sources, while at the same time increasing the water supply to the region during periods when there are seasonal fluctuations of availability.

The potential use of water provided by two basins when combined corresponds to 24% of the total amount of water required by the Industrial Complex [36]. Apart from a reduction in the amount of water extracted, the studies revealed a reduction in expenditure on energy in water withdrawal, and effluent (and contaminated rain water) treatment and disposal.

Some improvements may continue to be made in the all companies after the end of each project with UFBA due in part to the ideas that emerged and became conceptual projects which were routinely entered in the portfolio of the company's projects. The auditing of the sources of effluents allows the companies to reduce generation by controlling the source of waste, one of the most important assumptions of CP. In most cases, the company was able to lower its effluent treatment costs, reduce raw materials and input flows resulting from the minimization of discharges and leaks, and ceases to pay burdensome fines for releasing high amounts of contaminants that exceed the levels permitted by the environment regulations. The method developed in projects with industry has been adapted and expanded to other sectors that make use of water. Several projects have been carried out by TECLIM in urban settings such as public buildings, an airport, shopping centers and homes (www.teclim.ufba. br).

The method is currently being improved and new instruments are in development in current projects such as: environmental modeling and simulation of industrial processes; calculation of environmental indicators and its uncertainty from reconciled mass balance; supervision and monitoring of water consumption through the virtual system AGUAPURA (www.teclim.ufba.br/aguapura) and evaluation of impacts of projects in company after termination.

CONCLUSIONS

The TECLIM/ UFBA team is mainly concerned with conducting detailed investigation which involves identifying, analyzing and making recommendations on Industrial water systems, and thus breaking the pattern of inertia inside the industries regarding the close monitoring of water use.

The main results consist of the training of about 1700 industry professionals in Cleaner Production concepts, the identification of about 500 ideas of rationalization in water use, presentation and publication of 90 articles in journals, conferences and other academic events, identification of ideas with estimated potential savings of water of around 1400 $t \cdot h^{-1}$ and reducing effluent by at least 500 $t \cdot h^{-1}$. The instruments developed and presented here have led to the expansion of the method to home water reuse and urban issues that are also under consideration by the TECLIM Network.

ACKNOWLEDGEMENTS

The authors gratefully acknowledge the financial support given by FINEP (State Agency for Innovation) and the industries for the financial and technical support. Special thanks go to all the researchers which were part of the project's teams.

REFERENCES

1. A. Kiperstok, R. Kalid, K. P. Oliveira-Esquerre, E. Sales and G. Oliveira, "Rationalizing the Use of Water in Industry—Part 1: Summary of the Instruments Developed by the Clean Technology Network in the State of Bahia and Main Results Obtained," Journal of Environmental Protection, 2013.
2. A. Kiperstok, A. H. Tanimoto, D. Fontana, E. H. B. C. Silva, J. Mendonça, L. P. Lacerda, L. Pustilnik, L. F. Cardoso, R. Kalid and A. Teixeira, "The Fundamental Principles of Clean Production. Silver House: Introducing Clean Production in Bahia," Clean Technology Network (TECLIM), Federal University of Bahia (UFBA), Bahia, 2008, 446 p.
3. A. Kiperstok, G. L. Oliveira, K. P. Oliveira-Esquerre and R. Kalid, "Conservation of Water Resources in Semi-Arid Region Compared to Industrial Development," In: S. S. Medeiros, H. R. Gheyi, C. O. Galvão and V. P. S. Paz, Eds., Hydric Resources in Arid and Semi-Arid Regions, National Institute of Semiarid, Campina Grande, 2011, pp. 207-247.
4. K. P. Oliveira-Esquerre, A. Kiperstok, E. Cohim, R. Kalid, E. A. Sales and V. M. Pires, "Taking Advantage of Storm and Waste Water Retention Basins as Part of Water Use Minimization in Industrial Sites," Resources, Conservation and Recycling, Vol. 55, No. 3, 2011, pp. 316- 324. doi:10.1016/j.resconrec.2010.10.004
5. M. Martins, C. Amaro, L. Souza, R. Kalid and A. Kiperstok, "New Objective Function to Data Reconciliation of Water Balance," Journal of Cleaner Production, Vol. 18, No. 12, 2010, pp. 1184-1189. doi:10.1016/j.jclepro.2010.03.014
6. L. L. Massay and S. J. Udoka, "Industry-University Partnerships: A Model for Engineering Education in the 21st Century,"

Computers & Industrial Engineering, Vol. 29, No. 1-4, 1995, pp. 77-81.

7. R. M. Davies, "Industry-University Collaborations: A Necessity for the Future," Journal of Dentistry, Vol. 4, No. 1-2, 1996, pp. 3-5. doi:10.1016/0300-5712(95)00005-4

8. T. Barnes, I. Pashby and A. Gibbons, "Industry Interaction: A Multi-Case Evaluation of Collaborative R&D Projects," European Management Journal, Vol. 20, No. 3, 2001, pp. 272-285. doi:10.1016/S0263-2373(02)00044-0

9. J. Bruneel, P. D'Este and A. Salter, "Investigating the Factors That Diminish the Barriers to University-Industry Collaboration," Research Policy, Vol. 39, No. 7, 2010, pp. 858-868.doi:10.1016/j.respol.2010.03.006

10. J. Venselaar, "Environmental Training: Industrial Needs," Journal of Cleaner Production, Vol. 3, No. 1-2, 1995, pp. 9-12. doi:10.1016/0959-6526(95)00030-I

11. S. Herat, "Education and Training for Cleaner Production; a Feasible Learning Approach," Journal of Cleaner Production, Vol. 8, No. 5, 2000, pp. 361-364. doi:10.1016/S0959-6526(00)00038-X

12. T. E. Graedel and B. R. Allenby, "Industrial Ecology," Prentice Hall, New Jersey, 1995.

13. FIESP—Federation of Industries of the State of São Paulo, "Water Use and Conservation: Manual to Orient Industrial Sector," 2009. http://www.fiesp.com.br/publicacoes/pdf/ambiente/reuso.pdf

14. J. A. Romagnoli and M. C. Sánchez, "Data Processing and Reconciliation for Chemical Process Operations," Academic Press Inc., San Diego, 2000.

15. D. Fontana, R. Kalid, A. Kiperstok and M. A. S. Silva, "Methodology for Wastewater Minimization in Industries in the Petrochemical Complex," Proceedings of the 2nd Mercosur Congress on Chemical Engineering and 4th Mercosur Congress on Process Systems Engineering, Rio de Janeiro, 14-18 August 2004.

16. L. S. de Souza, "Propagation of Uncertainty in the Data Reconciliation with Linear Constraints," Monograph (Specialization in Industrial Automation with Emphasis in Computer Science, Instrumentation, Control and Optimization of Continuous Processes), Federal University of Bahia, Salvador, 2011.
17. L. S. Souza, R. A. Kalid, M. A. F. Martins, A. Kiperstok, K. Oliveira-Esquerre and L. Queiroz, "Propagation of Uncertainties in Data Reconciliation Without Redundancy Measurement," The 6th Brazilian Congress of Metrology, Natal, 27-30 September 2011.
18. R. Batavia, "Front-End Loading for Life Cycle Success," Offshore Technology Conference, Texas, 30 April-3 May 2001.
19. L. Saputelli, L. Lujan, L. Garibaldi, J. Smyth, A. Ungredda, J. Rodriguez and A. S. Cullick, "How Integrated Field Studies Help Asset Team Make Optimal Field Development Decisions," Proceeding of the Society of Petroleum Engineering, 31 March-02 April 2008.
20. C. E. Nobel and D. T. Allen, "Using Geographic Information Systems (GIS) in Industrial Water Reuse Modeling," Chemical Engineering Research & Design, Vol. 78, No. 4, 2000, pp. 295-303.
21. M. M. El-Halwagi and V. Manousiouthakis, "Synthesis of Mass Exchange Networks," General & Introductory Chemical Engineering, Vol. 35, No. 8, 1989, pp. 1233-1244. doi:10.1002/aic.690350802
22. K. P. Papalexandri, E. N. Pistikopoulos and C. A. Floudas, "Mass Exchange Networks for Waste Minimization: A Simultaneous Approach," Chemical Engineering Research & Design, Vol. 72, No. A3, 1994, pp. 279-294.
23. P. N. Sharratt and A. Kiperstok, "Environmental Optimisation of Releases from Industrial Sites into a Linear Receiving Body," Computers & Chemical Engineering, Vol. 20, Suppl. 2, 1996, pp. 1413-1418. doi:10.1016/0098-1354(96)00242-6

24. A. A. Ulson de Souza, H. L. Forgiarini, M. F. Xavier, F. L. P. Pessoa and S. M. A. Guelli de Souza, "Application of Water Source Diagram (WSD) Method for the Reduction of Water Consumption in Petroleum Refineries," Resources, Conservation and Recycling, Vol. 53, No. 3, 2009, pp. 149-154. doi:10.1016/j.resconrec.2008.11.002
25. A. Alva-Argáez, A. C. Kokossis and R. Smith, "The DeSign of Water-Using Systems in Petroleum Refining Using a Water-Pinch Decomposition," Chemical Engineering Journal, Vol. 128, No. 1, 2007, pp. 33-46. doi:10.1016/j.cej.2006.10.001
26. R. Smith and E. Petela, "Water Minimisation in the ProcEss Industries. Parts 5: Utility Waste," The Chemical Engineer, Vol. 523, Part 5, 1992, pp. 32-35.
27. Y. P. Wang and R. Smith, "Wastewater Minimisation," Chemical Engineering Science, Vol. 49, No. 7, 1994, pp. 981-1006.
28. J. A. C. Ruiz, "Decision Support Tools for Environmental Conscious Chemical Process," Thesis, Massachusetts Institute of Technology, Cambridge, 2000.
29. F. L. P. Pessoa, "Water Source Diagram. Course Handout on Methodology for Optimizing Environmental TECLIM. Tools to Minimize Water Use in an Industrial Environment," Federal University of Bahia, Salvador, 2008.
30. R. C. Moreira, "Reduction of Effluents in the Industrial Water Treatment Unit," Master's Dissertation, Federal University of Bahia, Salvador, 2009.
31. E. M. Queiroz and F. L. P. Pessoa, "Water Source Diagram Procedure: Wastewater Reduction for Single Contaminant," 17th International Congress of Chemical and Process Engineering, Praga, 27-31 August 2006.
32. S. F. Ávila, "Methodology for Minimizing Waste at Source from the Investigation of Abnormal Operating," Master Dissertation, Federal University of Bahia, Salvador, 2004.

33. S. F. Ávila, "Control Processes in the Generation of Organic Effluent," 35th Brazilian Congress of Chemical Engineering, Salvador, 1995.
34. K. P. Oliveira-Esquerre, A. Kiperstok, R. A. Kalid, E. Sales, L. Teixeira and V. M. Pires, "Water and Wastewater Management in a Petrochemical Raw Material Industry," 10th International Symposium on Process Engineering, Salvador, 16-20 August 2009, pp. 1047-1052.
35. K. P. Oliveira-Esquerre, A. Kiperstok, P. R. C, Penalva, R. Kalid and P. P. Quirino, "Study of the Potential for the Reuse of Exploited Water by the Hydraulic Barrier System of Camaçari Industrial Pole of Bahia," 8th Symposium Luso-Brazil of Sanitation and Environmental Engineering, Silubesa Belém, 10-14 March 2008.
36. K. P. Oliveira-Esquerre, A. Kiperstok, E. Cohim, R. Kalid, E. A. Sales and V. M. Pires, "Taking Advantage of Storm and Waste Water Retention Basins as Part of Water Use Minimization in Industrial Sites," Resources, Conservation and Recycling, Vol. 55, No. 3, 2011, pp. 316-324. doi:10.1016/j.resconrec.2010.10.004